Cultural Tourism in the Asia Pacific

Tai-Chee Wong • Hoon-Peow See
Meg Milligan
Editors

Cultural Tourism in the Asia Pacific

Heritage, City and Rural Hospitality

Springer

Editors
Tai-Chee Wong
New Era University College
Kajang, Selangor, Malaysia

Hoon-Peow See
Sunway TES College
Iskandar Puteri, Johor, Malaysia

Meg Milligan
Department of Psychology
Troy University
Troy, AL, USA

ISBN 978-3-031-63458-1 ISBN 978-3-031-63459-8 (eBook)
https://doi.org/10.1007/978-3-031-63459-8

This Springer imprint is published by the registered company Springer Nature Switzerland AG
The registered company address is: Gewerbestrasse 11, 6330 Cham, Switzerland

Foreword

Cultural tourism is expanding rapidly in Southeast Asia, supported by government initiatives and commercial opportunities. It has survived the COVID pandemic, geopolitical concerns, and economic turbulence. Onsite tourism by individuals and groups is expanding even as virtual tours of many of the same venues are increasingly available via satellite and the Internet. Developments in technology and communication have been a boost, not a hindrance, to the development of cultural tourism.

So much about this topic is new, especially the expansion in its scale and scope. Yet it is also a continuation of the centuries-old pattern of travel, for so long prominent in Western cultures, for personal development and maturity. Yet, as the editors advise, its roots date to at least the 7th century in Asia.

The function of this book is not simply to extoll the many wonderful venues in Asia, nor of specific travel sites. Neither is it a manual of how-to advise for travelers. Nor does it merely describe and elucidate emerging trends, for example, the increasing travel among Asian nations. Rather, it concerns some of the myriad problems and issues in this rapidly growing arena, along with emerging solutions, exemplified in site-specific cases. This is indeed a book about issues. It deals with purposeful change, as well as recognition and response to unanticipated change. It concerns problems of both government and commercial administration. Cases in the book concern conflict, and the satisfaction of divergent stakeholder needs. In addressing this challenge, the authors employ a variety of pure and hybrid methodologies. They sometimes adapt well-known models or metrics or devise their own for the case in question. It is a work of value for administrators, academics, and of course informed tourists. Several of the case studies describe successful efforts to develop cultural tourism around specific cultures and highlighted practice of their heritage.

The editors and authors in this work are well acquainted with Southeast Asian travel and can bring their professional and academic expertise to bear on the many issues. I became acquainted with editors, especially Dr. Milligan and Dr. See, through our collaborative studies on educational leadership in Asia. They are individuals of dogged diligence, also willing to tackle projects in diverse subject areas. Working with them has been a valuable and meaningful experience.

Singapore and Malaysia are especially noted for their ethnic diversity, with large populations of ethnic Malays, Chinese, and Indians. Local attractions and amenities reflect this richness of multiple heritages. Dr. Tai-Chee Wong describes the evolution of the "product" provided by cultural tourism in Singapore. Since the tourist destinations are now primarily urban and provided to larger groups, there may be a more and more diverse number of salient, legitimate stakeholders. The quality attributes of the tourist experience have thus changed considerably over time, as have the vehicles for providing the experience. The trend has been toward higher end experiences, with greater variety as well.

Using Malaysian sites and tales as examples, Dr. Hoon-Peow See provides a charming discussion of the value, promotion, and consumer experience of folklore, the oral traditions of tales, myths, legends, costume, dance, art, etc. Each cultural site may have its own folklore. Where different ethnicities share a region, each may have its own folklore. Nevertheless, offsite techniques, such as cultural shows or online experiences may prove valuable for promotion and enhancing the experience.

Dr. Meg Milligan and her three colleagues provide an excellent discussion of the inclusion of people with disabilities (PWD) in cultural tourism. This is promoted by both national mandates and international arrangements such as the United Nations Convention of Rights of Persons with Disabilities and the International Classification of Functioning, Disability, and Health. The authors review the relevant legal and practical situation in Asia, with special attention to Malaysia. They then report their onsite study of accessibility at Kinabulu Park in Malaysia. In another Malaysian study, Tai-Chee Wong considers the optimal utilization of natural amenities, with Sabah National Park as its focus. Calculating the optimal number of park admissions involves a tradeoff between revenues and site degradation through overuse. The study invokes ecological concepts and economic analysis.

Cultural tourism is especially attractive for many countries, and especially underdeveloped areas with scarce capital but relevant amenities. Building on accepted work on tourism development, James Faumuina synthesized an experiential cultural tourism model (ECTM) for analysis of economic problems of the new nation states of Micronesia, Marshall Islands, and Palau, with specific application to the recovery of Bikini Atoll into a tourism and educational site. In a contrasting paper, Dr. See and Dr. Wong describe the general aspects of "edutourism" in the Asia Pacific region. Although it has ancient roots in Asia, this form of tourism has exploded in popularity due to rising affluence and ease of travel, resulting in an astounding array of degree and certificate programs, internships, workshops, and the like.

Some of the case studies concern cultural sites in mainland China, with the attendant resources, challenges of this venue. To succeed in the centralized structures of China, local champions of tourist sites must organize well and also influence the party and senior government officials. In a case study of the tensions in government promotion of "Red Tourism" to historical sites of the revolution, Dr. Liu analyzes the case of tourism promotion at Taopeng, a mountain village in rural Beijing that served as a base of resistance to the Japanese invasion. In promoting the project, local leaders courted government and private involvements and financing in developing the site, the content, and the necessary tourist infrastructure. They advertised heavily toward youth, using digital media and analysis with artificial intelligence tools. The tours are partly re-enactments, utilizing the clothing and symbols of the revolutionary period. Nevertheless, there remains a conflict between the revolutionary values expounded and the consumerist ambience of the site.

In a charming discussion focusing on the heritage sites of Malaysia, Dr. See explores the problems of developing, promoting, and improving tourism that centers on folklore—the oral tradition of tales, music, and myth—as a sustainable product.

Much of cultural tourism is based around local, traditional ethnic festivals, which embody the core of local cultures. Tourists seek these events especially. Dr. Lei emphasizes that such festivals display cultural symbols and embody spiritual beliefs. Organization of these festivals on a commercially feasible scale, often by government, may omit significant details and nuances. Personal interactions at these events are a two-way transfer, as local inhabitants are also exposed to modern elements. Her contribution considers the issues for balancing promotion of meaningful tourism with preservation of local culture, training of young organizers and participants, careful inclusion of rituals and artifacts, and development of local museums. Modern elements and commercial practices should be encouraged, but within these constraints.

Thailand has impressive scenery, rich history, political openness, and a number of UNESCO heritage sites. It has invested heavily to promote these in cultural tourism. Dr. Hussain and his colleagues provide a strong overview of these amenities and the national and local efforts to preserve and promote them.

After severe political unrest and shocks from COVID, Hong Kong reopened as a tourist venue in late 2022. As an urban venue, lacking many of the popular, spectacular natural features, Hong Kong relies more on truly cultural and artistic resources. Professor Lim describes this reopening and in particular this Special Administrative Region's largest tourism project, the West Kowloon Cultural District (WKCD).

Mr Wang and his colleagues describe the evolution of training and presentation of the traditional Lion Dance, which "symbolizes the national soul of the Chinese nation," and dancers perform it every Chinese New Year. As a sort of "team sport," it fosters and inculcates national pride, teamwork, and perseverant efforts.

This is a well-curated book of well-written, informative articles. It is well worth reading for anyone wishing to travel to cultural sites in Asia.

Professor, Department of Management
Troy University
Troy, AL, USA

John M. Mankelwicz

Contents

About the Editors

Tai-Chee Wong has his BA and MA degrees in urban and regional planning from the University of Paris and earned his PhD in Human Geography from the Australian National University. After teaching 20 years at Nanyang Technological University, Singapore, he retired in 2013. He then worked as Distinguished Professor for two years at Guizhou University of Finance and Economics, China, and as Dean and Professor at the Southern University College, Johor Malaysia, from March 2017 to December 2018, and as Deputy Vice-Chancellor at Berjaya University College, Kuala Lumpur from October 2022 to September 2023. He was a visiting professor at the University of Paris (Sorbonne IV) and a visiting fellow at Peking University and Tokyo University. His main research interests are in urban and economic issues and, more recently, in Malaysian politics. Besides his 16 self-authored and edited book volumes, he has written over 100 academic articles and published widely in international journals including tourism papers.

Hoon-Peow See holds a PhD in Sociology and Anthropology from the International Islamic University Malaysia, a Master of Philosophy in Sociology from the University of Malaya, and an LLB (Hons) from the University of Glamorgan (now the University of South Wales). He has held various leadership positions in the education industry with exposure in both academic and business operations, which include being CEO cum Vice Chancellor of a university college. He is currently the Director at Sunway TES College Johor Bahru, Malaysia. He has a wide academic interest in humanities and social sciences. Although he is a full-time manager, he is active in research and publishing. He has published many books and papers locally and internationally with reputable publishers and journals. He has also presented papers at local and international conferences.

Meg Milligan, PhD in Sociology and Anthropology, American University of Beirut, Lebanon, PhD in Counseling Psychology, Auburn University, Alabama, USA, psychology internship at the Veterans Administration Medical Center, Tuskegee, Alabama, licensed psychologist in Alabama, is a professor in the Department of Psychology and adjunct faculty in the global leadership doctoral program at Troy

University, Alabama. Milligan has thirty years of teaching experience in higher education in Malaysia (2016–2022) and the United States; conducts study abroad programs annually; has presented over 100 papers and posters at conferences in nine countries spanning three continents; has faculty, national research, and international research excellence awards; and has publications on disability rights, narcissistic leadership, and global identity, among others. Milligan is President-Elect (2023–2024) of the Association for Distance Education and Independent Learning (ADEIL) after two terms on their Board of Directors, will be President (2024–2025), and has Quality Matters (QM©) master reviewer certification at all levels.

Chapter 1
Cultural Tourism in the Asia Pacific: An Introduction to Its Heritage, City and Rural Hospitality

Tai-Chee Wong, Hoon-Peow See, and Meg Milligan

1 Background of Cultural Tourism in the Asia Pacific

From the 1960s, soon after the start of jet-propelled chartered flights, expansion of tourism has been extended from the affluent West to the developing world. The latter has since the 1990s benefited from the budget air whose publicity is "everyone can travel". Soon after, niche areas of tourism like heritage tourism, arts tourism, gastronomic tourism, film tourism and creative tourism have also made their appearance. More recently, the scope has been extended to as far as to cover vacation as public health resource where wellness-centred tourism has been designed (Lehto & Lehto, 2019). With the regulatory support of the United Nations, large numbers of national governmental tourism administrations accepted cultural tourism as part of their tourism policy, a sector seen as a substantial income generator (Richards, 2019).

By 2019, prior to the worldwide spread of the Covid-19 which had incurred great damages temporarily to the tourism industry for 4 years, there were an estimated 1.5 billion international arrivals. With projected growth at the wake of the Covid-19 epidemic, improved connectivity for destinations, technological innovations, and the stronger urge to move towards a more sustainable "mother earth" and inclusive growth by incorporating benefits to the populace, dramatic transformation is expected to take place within tourism itself as the largest service sector in the world (OECD, 2020).

T.-C. Wong (✉)
New Era University College, Kajang, Selangor, Malaysia

H.-P. See
Sunway TES College, Iskandar Puteri, Johor, Malaysia

M. Milligan
Department of Psychology, Troy University, Troy, AL, USA
e-mail: mmilligan@troy.edu

T.-C. Wong et al. (eds.), *Cultural Tourism in the Asia Pacific*,
https://doi.org/10.1007/978-3-031-63459-8_1

Indeed, growth of cultural tourism, like the consumption of automation-produced manufactured goods, is underlain by rising numbers of middle classes and ease of travel and affordability. With automation and Fordist system of production which has helped mass-produce standardized goods, industrial products have been transformed into reasonably priced goods fit for mass consumption (Bolz, 2011). Almost by the same token, tourism has gradually developed from a small niche market to a gigantic consumption market generating substantial amount of revenues and jobs. Visits to cultural and historic sites, nature attractions and conventional events, have turned out to be a specific form of consumption which has an increasingly strong appeal to travelers or leisure seekers.

As a result of the rapid growth of tourism and job prospects, especially in the emerging markets, tourism and hospitality has been incorporated as a widely accepted teaching discipline in their tertiary institutions where scholars have published numerous quality articles in reputable journals and books, whether self-authored or edited volumes. In this introduction, we wish to cite a few well received book volumes as a back-up reference underlying the essence of our edited volume on cultural tourism in the Asia Pacific.

Kaye Sung Chon and Hilary du Cros (2002) published their well conceived book by Routledge titled *Cultural Tourism: The Partnership Between Tourism and Cultural Heritage Management.* A key issue examined in this book is the conflicting relationship between tourism business professionals who evaluate tangible (physical assets of culture) and intangible assets (ongoing cultural practices, knowledge, and living experiences) from business and profit making approach, and those heritage professionals who judge the same assets by assessing their intrinsic value on the basis of sustainability and need for conservation. The true partnership therefore is the compromise between highly economic pursuits and protective measures needed for sustainable development. A model is ultimately reached between the two authors, one tourism specialist versus the other cultural heritage management expert where they integrate management theory and practice from two diverse perspectives in a working partnership expected to offer mutual benefits to both.

Dallen Timothy's (2020) book by Channel View, "Cultural Heritage and Tourism: An Introduction", provides a rich theoretical overview and content on tourism practices, notions of proper management issues on cultural heritage tourism. Timothy has also dealt with broader issues such as geopolitical conflicts, solidarity tourism, social media and indigenous knowledge that could have led to varied impacts on heritage tourism. More interestingly, his concerns on excessive tourism, climate change, have inevitably impacted on environmental change, heritage economics in heritage sites, among others.

Another volume edited by Pei-Lin Yu, Thanik Lertcharnrit and George Smith (Yu et al., 2023), *Heritage and Cultural Heritage Tourism: International Perspectives,* published by Springer, portrays a rich collection of essays presenting cultural heritage and tourism issues globally. The key themes dealt with cover a wide spectrum including historical and economic analysis of cultural tourism in selected societies and communities, heritage sites are studied as a source of social and cultural capital and their being explored in depth with challenges and opportunities. There are cases being studied in the sustainability context of cultural heritage, with highlights in particular on East and Southeast Asia.

2 Key Concerns of Cultural Tourism

As cultural tourism grows further and becomes more popular to the masses, the importance of cultural tourism is obviously imminent to both national governments and ordinary travelers alike. With an estimate of the total of 1.8 billion international tourists by 2030 across the globe, the number will be even more if domestic arrivals are included. In a vast and populous country like China, for example, its domestic tourist figure in 2019 was more than 20 times higher than its outbound tourists. In 2019, China's outbound tourists registered merely 62.88 million but its growth potential is expected to be high in the near future (CEIC, 2021a, b).

Cultural tourism researchers may have different focuses and orientations, they do however share to a great extent common concerns in their approach. These concerns have been broadly addressed by the Organization for Economic Cooperation and Development (OECD, 2020) which we will briefly discuss below.

Primarily, one of the priority concerns is the sustainable growth of cultural tourism in the perspective of the economic, social and environmental benefits and costs. Whether individual destinations could achieve a balanced economically, socially and environmentally development, it relies heavily on appropriate measures vis-a-vis the challenges and unique characteristics they each face as there is no "one size fits all" solution. In this aspect, to prevent over-congestion leading to adverse environmental impact, those destinations experiencing rapid and unplanned growth must pay particular attention and care to their sensitive natural and cultural sites to cope with rising visitor numbers.

By principles, measurement of tourism success should not be built solely on numbers but on the accountability of whether it sustains in the long-term interest of the sites and its ability to contribute to local economies and the community. Local communities' indigenous traditions, art, and culture, which are important sources of tourist attractions will only be consolidated and rationally maintained with upgraded transport infrastructure and improvement of their quality of life. This improvement is only achievable in real terms with funding regenerated from tourism benefits with strong political leadership and support of all relevant stakeholders (OECD, 2020: 90–91).

Secondly, what is seen as utmost important in the preservation of cultural tourism is the need to promote cultural authenticity and natural assets and this must be constructed with the involvement of stakeholders and sustainable development principles (OECD, 2020: 91). By principles, we mean the necessity to raise awareness of the business operators, visitors and public agencies with regard to the cultural and environmental values towards the protection and management of protected areas and sensitive sites. Simply said, revenues generated from tourism activities should be redirected in an appropriate way towards a double objective of destinations: environmental conservation, preservation of cultural heritage and its authenticity.

The third concern applies more to over-growth in unplanned urban and nature-based tourism as a result of inappropriate tourist behaviour, rising noise pollution and pedestrian and vehicular traffic, and the degradation of community-specific infrastructure and facilities (OECD, 2020: 93). Over-commodification of cultural events can result in the loss of authenticity and cultural identity and heritage, given the pressure in the need in business to make adjustments to meet attitudes and values of tourists from a diversity of cultural backgrounds. Consequently, an over-dependence on tourism income is liable to the loss or weakening of cultural traditions and values.

3 Objectives of the Book

In this proposed book volume, the conceptualization of cultural tourism and hospitality is dealt with in a broad sense. The types of papers we call for can thus include culture-based conceptual papers, empirical case study writings, and those from the industry players who could integrate their rich on-the-job experience into a topical tourism research. They all fall within an urban or rural spatial setting.

At a more investigative level, cultural tourism covers a wide aspect of studies inclusive of tangible and intangible cultural heritage. Whilst tangible elements are visible physical objects, intangible cultural heritage elements are much more complex, unique and deeply culture-rooted in different societies. These elements comprise non-visible internal and dynamic mechanisms inherent in their historical, cultural and socio-economic past experience and value systems that have been passed down from generation to generation. UNESCO (2003: 3) highlights the following four key components in its "Convention for the Safeguarding of the Intangible Cultural Heritage":

(a) Traditional, contemporary and living at the same time: it represents inherited traditions and encompasses contemporary rural and urban practices of a diversity of cultural groups;
(b) Inclusivity of cultural expressions: expressions of intangible cultural heritage may be shared among different social groups. But expressions might have been adapted over time by one group or another in response to their environments, giving each a sense of identity and continuity which contributes to social cohesion, encourages a sense of identity and responsibility;
(c) Representativeness: it is exclusive or has its exceptional value; and
(d) Community-based: it is recognized as such by the communities, groups or individuals that create, maintain and transmit it.

Within these elements, there are distinctive sets of intellectual, spiritual and emotional features which are representative of a wide variety of societies having their own respective architectural arts, literature, music and living lifestyles. Each of them all possesses a character that merits careful survey and in-depth study. In this regard, one has to bear in mind that a careful differentiation is required when cultural transmission is shifted from the West to the Asia Pacific whereby it has its own evolutionary insertion of Asian values and context (see You & Hardwick, 2020). Adaptation and adjustment are thus seen as always necessary.

As a matter of fact, cultural heritage is one which is powerful in inducing travel interest and it has great potential to protect and conserve historic sites sustainably given its viability in income generation if carefully designed and managed. Gastronomy culture which is unique from place to place characterized by its diversity, is another attractive heritage that promotes not only consumption of daily needs while travelling but also cultural interchange of people at cross points around different world regions.

With cultural tourism which is expected to continue to grow in both business and research, the present editors of this proposed volume have hope that they will be able to make further contributions in this domain by exploring the Asian field context where there is still much to be uncovered.

This book is a collection of essays on cultural tourism which has a very broad coverage in conceptual interpretations and practices, with a particular focus on the Asia Pacific region. It is widely agreed that cultural tourism has a strong linkage with tourist-based activities in urban, nature heritage, socio-cultural heritage, among others. These activities are again interlinked with their sustainability values in environmental, economic and social aspects out of the need to preserve them in the long-term, yet having them to act as leisure destinations with potential in generating revenues and jobs.

In the above connection, China, for instance, since signing the World Heritage Convention in 1985, had increased by July 2021 its intangible world heritage sites to 56, and its natural heritage sites to 14. It is believed that China has sought a series of innovative means to transform its cultural heritage sites into high-quality global cultural heritage tourism destinations while assuming responsibility to protect them. However, a problem has remained that the Chinese government has yet to streamline its administration measures nation-wide and delegate power to local authorities to protect the cultural heritage (Zhang et al., 2023; Timothy et al., 2009).

On the domain of nature heritage conservation, the introduction of tourism activities into a protected nature area, must not be conducted at the expense of those of locals, destruction of the regeneration capacity of indigenous plant and animal species. In this sense, it is imperative that companies or individuals running tourism business in such a protected zone should avoid taking advantage of the country's "existing situation of poverty, corruption and weak law enforcement" to gain maximal profits (Labadi, 2022: 188).

One of the key features of cultural tourism is the pursuit of its sustainability values. By sustainability, we mean not merely the preservation of physical objects *per se* but the conservation of its physical environment, plus other socio-economic activities that are critical to sustain them in the long term as well. To achieve sustainability, there is also a need to address the technocratic and ecocentric approaches. By technocentric approach, sustainable development issues are quantified and solved largely with the help of technological measures whereas ecocentric approach asks for an assurance of quality of life which is seen as even more important than economic growth values (Mason, 2016; Holden, 2000).

Conceptually but perhaps over idealistically in the real world faced with rising numbers of tourists, Mason's (2016: 215, 217) idea of sustainable tourism involves the notion of deep ecology where emphasis is placed on the preference of low-scale technology, and moral respect of the "rights of nature" to be conserved because the environment possesses an intrinsic value that outweighs its value as a tourism asset.

4 About the Chapters

The listing of the chapters in this book volume follows the logical sequence from the regional perspective first and then turn to country to country in the Asia Pacific. Following Chap. 1 which is the Introductory Chapter, we deal with educational tourism issues in Chap. 2 which covers countries across the Asia Pacific. In recent decades, we notice the rise of Asia Pacific countries notably Japan, China and Malaysia as popular destinations in higher education and short-term programs for international students. Standard-wise, Asian countries are catching up with their counterparts in the West as a provider for quality higher education to international students. Beside regular tertiary programs, there are other business-oriented, practical-based and professional training programs such as student exchange, language immersion, volunteer, cultural/historical/environmental tours, workshops and internship. They cover a wide variety of topics and contents, and they last from a couple of weeks in one program to as long as six-months or longer in another. Aiming at international students who are required to pay high fees, educational tourism has helped bolster national revenues and in successful cases, build up a nation's international educational reputation.

Tai-Chee Wong's Chap. 3 discusses "Tourism Development of Singapore: Artificiality versus Profitability". As a pragmatic government, Singapore views the tourism sector as one of its key service sectors to create high economic value and benefits. For lack of historic heritage and much of this has also been phased out in its modernization process, Singapore has created a series of urban tourism attractions to beautify its scenic tropical environment from large convention centres to resort, bay gardens and casinos. The author examines the success and weaknesses of cultural tourism being put in place within such an artificial setting where man-made features complement genuine heritage attractions. This combination is economically a success, though in some cases artificiality may be misjudged and incompatible in its place remaking or reimaging process as there could have created missing links in historical and cultural continuance. Despite certain negative impacts, Singapore's urban tourism has enhanced quality of life, characterized by the rise in local incomes and improved public transport. Moreover, Singapore tourist agencies are resilient and are highly adjustable to fit in new circumstances to counter challenges.

In Chap. 4, Meg Milligan, Lew-Lee Ting, Suzanne Tew-Washburn, and Mary Kristin Medina's chapter, "Inclusion in Tourism: Cultural heritage sites in Malaysia" examines the imperative for cultural heritage sites to be available to all persons, including persons with disabilities, the largest minority group in the world. People with disabilities are an underserved tourist group, which presents growth opportunities for the tourism industry. The authors present a developmental history of perspectives, terminology, guidelines, and laws related to inclusion and, ultimately, universal design from global, such as the World Health Organization and the United Nations, to national depictions, including Malaysia. Kinabalu National Park, a United Nations Educational, Scientific, and Cultural Organization (UNESCO) World Heritage Site in Malaysia, is presented as an example of accessibility and inclusion in a cultural heritage site. In addition, this location illuminates challenges inherent in accommodating neuro-divergent tourists in nature-based sites. Addressing these concerns has moral, social, economic, and legal benefits for the tourism industry and its stakeholders.

Hoon-Peow See's Chap. 5 titled "Folklore for Cultural Tourism: The Malaysian Experience" investigates the value of heritage sites and cultural activities/performances which has been long discovered to have tourism appeals. However, the value of folk narrative or the intangible/oral heritages, such as local folk tales, legends and myths, which are very much a part of the local culture, have not been fully realized in Malaysia. This study explores their inherent values in two aspects: attracting tourists through promotion and improving tourists' experiences, using Malaysian sites and tales as examples. The key approach is by mainly verbalizing or visualizing the intangible folklore, thus making them part of the travel experience. Public agencies such as Tourism Malaysia and the Ministry of Tourism Malaysia have made good use of Malaysian myths and legends to promote tourism since the 1950s, but their approach is not without weaknesses. Firstly, the stories selected are mostly old Malay or indigenous tales, with very few related to colonial history, and the use of non-Malay folklore is limited.

Secondly, facilities like legend parks, LED lighting trails, museums at some legendary sites have not been well-maintained, thus affecting their presentation to visitors. More cost friendly alternative approaches to promote folklore such as cultural shows, story telling, multimedia shows, murals etc., should be explored. Thirdly, local guides and other tourism professionals should be educated about these tales to share with visitors.

Tai-Chee Wong's "Examining the Trade-offs of Visiting Fee Rises to National Parks, Sabah, Malaysia" in Chap. 6 deals with a controversy over admission fee hikes and its impact on visitor numbers. Apart from this, there are other debatable issues associated with fee hikes. This chapter examines the trade-off effects resulting from fee hikes with reference to the issues of carrying capacity, environmental justice and governance as the basis whether visitors should be restricted in numbers, justified by the need for nature conservation by cutting down ecological footprints. There is a tension between conservation and tourism development goals and the inherently complex conflict of interests among stakeholders. To minimize the footprint impact, there is a need to include appropriate policy design of environmental conservation, considerate tourists highly conscious of their duty to protect the park. Conflict resolution should be a balance between scientific research findings about conservation and political consensus involving stakeholders inclusive of policy makers, management agency and park users.

Chapter 7 by Kashif Hussain and his team is titled "Thai Cultural Tourism Attributes: Emerging Trends and Sustainable Practices" and it highlights the ethical perspective that advocates the need to satisfy tourist demand, without compromising the long-term sustainability of the Thai cultural tourism heritage. Key conservation sites include the World Heritage Sukhothai and Ayutthaya, as well as the famous Thai beaches like Pattaya and Phuket. To date, Thailand has switched on to a sustainable approach manifested in legislative measures, green programs, and local engagement. Cultural premises are varied in Thailand and these include religious temples, arts and cultural centres as well as culinary traditions. Ecotourism and adventure travel tours which allow exploration of Thailand's abundant natural biodiversity is also getting popular. The chapter discusses issues related to the over-tourism with a focus on balancing tourist demand with the benefits of local community and cultural and natural heritage protection. Investments in cultural infrastructure, preservation projects and technological innovations for an upgraded tourism experience are highlighted. Future pathway includes addressing over-tourism challenges through striking the balance between economic growth and cultural conservation and developing public-private partnership for sustainable cultural tourism in Thailand.

In Chap. 8, Benny Lim seeks to explore the strategies that drive West Kowloon Cultural District (WKCD) in Hong Kong as a cultural tourism destination. Over the span of 3 years, four major cultural institutions, Xiqu Centre in early 2019, Freespace in mid-2019, followed by M+ Museum (M+) and the Hong Kong Palace Museum (HKPM) in November 2021 and July 2022 were launched respectively. Cultural districts provide significant cultural value to its residents and local communities by allowing them to develop their "cultural capital". In Lim's own words "Cultural districts enrich this cultural capital by providing a diverse range of cultural and artistic experiences such as art exhibitions, theatre performances, concerts, and workshops, allowing individuals to broaden their cultural quotient, increase their understanding and appreciation of various art forms, and engage in creative expressions". Besides, various strategies targeted at the WKCD has been adopted to promote its functional operations, and Benny Lim has also pointed out its untapped potentials to be developed into a more significant destination where effective involvement of stakeholders is sought after.

Ran Liu's Chap. 9 is "Red tourism in Rural Beijing: The Hierarchical Governance and Grassroots Community Engagement". Red tourism is a top-down ideological infusion by the CCP into heritage tourism, largely rural-based in China, and it advocates the collectivism-oriented political consciousness on the one hand, and promotes a "consumer culture" on another. This chapter examines an unconventional star branding of "red heritage" and "red culture" in Beijing's red tourism villages. A case study is used to investigate how a deep-mountain and poorly equipped Taopeng Village with less than 100 rural households has been miraculously branded a shining star in the process. With the help of the local grassroots government, remote hilly villagers are mobilized and act as bottom-up creative entrepreneurs. Taopeng Village has employed various strategies to attract visitors, including the mobilization of public and private corporations and schools to organize trips; and for the young, a one-day immersive experience. There are however potential conflicts between the ideology of patriotic education and the logic of "consumerism" in the tourism consumption pattern. For instance, boutique homestays and leisure cafés are designed more for middle-class consumption demands rather than for the grassroots.

In Chap. 10, Xiangzhu Lei's "Thinking about Ethnic Minority's Rural Festival Cultural Tourism: A Case of Guangxi, China" has dealt with unique ethnic festivals in China's mountainous southwestern region. Here ethnic festivals play a dual role which embodies on one hand part of the core culture and identity of different ethnic groups, and at the same time a key source of tourism activities on which many make a livelihood. However, these festivals are increasingly intervened by the public authorities which often aim to accommodate large audiences by organizing large-scale performances in the form of folk parades, thus transforming the mysterious, solemn, and culturally distinctive rituals into a shallow performance. This has resulted in the loss of the original essence of the festivals. To counter the over-commercialization of festivals which has tended to threaten the authenticity and originality of traditional festivals, the author has sought to balance off such a trend with emphasis on training the younger generation to appreciate their own cultural heritage and ritual practices. Another approach recommended is to "blend" ethnic festivals with heritage sites and by doing so it consolidates the cultural connotation of each ethnic minority group with their past. This not only will help improve the quality of rural tourism but it also prevents the festivals from being disconnected with the genuine cultural sites.

Xiaoyu Wang's Chap. 11 called "Promoting Lion Dance as an Intangible Cultural Heritage Via Education in Lingnan, China" is a discussion of Lingnan Lion Dance as a cultural heritage in South China. Lion dance serves a dual role, manifesting both its unique charm as intangible cultural heritage and its popularity as a festive sports activity. In recent years, Lingnan Lion Dance has gained prominence and attracted the training of Lion Dance practitioners. Consequently, the training has shifted from the traditional apprentice system to professional training in vocational schools under the guidance of skilled coaches. Through an in-depth investigation into the teaching methods of specific schools, the author has discovered that courses oriented towards practical-based training have sparked students' interest and helped them comprehend better the cultural history of Lingnan Lion Dance. As a collective cultural sports activity, Lion dance instills values such as teamwork, unity, resilience, and mutual tolerance, contributing significantly to the transmission of Lingnan Lion Dance traditions. Public performances of Lion Dance have also greatly promoted it as a form of tourism activity in domestic and international festive events.

James Faumuina's Chap. 12, "Transformative Tourism: Introducing the Experiential Cultural Tourism Model as an Innovative Means to Cultivate Resilient

Ends for COFA Nations" introduces a new experiential cultural tourism model (ECTM). This model is based on a novel perspective that could transform the tourism sector and its impact globally. The author deconstructs the traditional tourism approach and proposes a way to distribute more equitably its benefits among "shareholders" while emphasizing community empowerment and the transformation of passive tourists into active participants in environmental restoration, sustainable practices, and positive social change that respect local culture. The author describes the development of his model and acknowledges the influences of The Resilient Intersectional Community-Based Empowerment (RICE) Framework and the Community-Based Economic Development Wealth Equity Building Blocks (CBED WEBB) concept. He uses the Compact of Free Association (COFA) nations (The Federated States of Micronesia, the Republic of the Marshall Islands, and the Republic of Palau) as an example of the practical utility of the ECTM.

References

Bolz, R. W. (2011). *Manufacturing automation management: A productivity handbook*. Chapman and Hall.

CEIC. (2021a). *China's domestic tourists 1994–2021*. https://www.ceicdata.com/zh-hans/china/outbound-tourist

CEIC. (2021b). *China's outbound tourists 2004–2021*. https://www.ceicdata.com/zh-hans/china/domestic-tourist

Chon, K. S., & Du Cros, H. (2002). *Cultural tourism: The partnership between tourism and cultural heritage management*. Routledge.

Holden, A. (2000). *Environment and tourism*. Routledge.

Labadi, S. (2022). *Rethinking heritage for sustainable development*. UCL Press.

Lehto, X. R. Y., & Lehto, M. R. (2019). Vacation as a public health resource: Toward a wellness centered topurism design approach. *Journal of Hospitality & Tourism Research, 43*(7), 935–960.

Mason, P. (2016). *Tourism impacts, planning and management* (3rd ed.). Routledge.

OECD. (2020). Chapter 3 in OECD Tourism trends and policies. In *Rethinking tourism success for sustainable growth* (pp. 89–122). OECD Publishing. https://doi.org/10.1787/82b46508-en

Richards, G. (2019). Cultural tourism: A review of recent research and trends. *Journal of Hospitality and Tourism Management, 36*, 12–21.

Timothy, D. J. (2020). *Cultural heritage and tourism: An introduction*. Channel View.

Timothy, T. J., Wu, B. H., & Luvsandavaajav, O. (2009). Heritage and tourism in East Asia's developing nations: Communist-socialist legacies and diverse cultural landscapes. In T. J. Timothy & G. P. Nyaupane (Eds.), *Cultural heritage and tourism in the developing world: A regional perspective* (pp. 93–108). Routledge.

UNESCO. (2003). *2003 Convention for the safeguarding of the intangible cultural heritage*. https://ich.unesco.org/doc/src/18440-EN.pdf

You, Z. Y., & Hardwick, P. A. (2020). Intangible cultural heritage in Asia: Traditions in transition. *Asian Ethnology, 79*(1), 3–19.

Yu, P. L., Lertcharnrit, T., & Smith, G. S. (Eds.). (2023). *Heritage and cultural heritage tourism: International perspectives*. Springer.

Zhang, S. B. W., Liang, J. X., Su, X. W., Chen, Y. C., & Wei, Q. (2023). Research on global cultural heritage tourism based on bibliometric analysis. *Heritage Science, 11*, 139. Open Access. https://heritagesciencejournal.springeropen.com/articles/10.1186/s40494-023-00981-w

Tai-Chee Wong has his BA and MA degrees in Urban and Regional Planning from the University of Paris and earned his PhD in Human Geography from the Australian National University. After teaching 20 years at Nanyang Technological University, Singapore, he retired in 2013. He then worked as Distinguished Professor for two years at Guizhou University of Finance and Economics, China, and as Dean and Professor at the Southern University College, Johor Malaysia, from March 2017 to December 2018, and as Deputy Vice-Chancellor at Berjaya University College, Kuala Lumpur from October 2022 to September 2023. He was a Visiting Professor at the University of Paris (Sorbonne IV) and a Visiting Fellow at Peking University and Tokyo University. His main research interests are in urban and economic issues and, more recently, in Malaysian politics. Besides his 16 self-authored and edited book volumes, he has written over 100 academic articles and published widely in international journals including tourism papers.

Hoon-Peow See holds a PhD in Sociology and Anthropology from the International Islamic University Malaysia, a Master of Philosophy in Sociology from the University of Malaya, and an LLB (Hons) from the University of Glamorgan (now the University of South Wales). He has held various leadership positions in the education industry with exposure in both academic and business operations, which include being CEO cum Vice Chancellor of a university college. He is currently the Director at Sunway TES College Johor Bahru, Malaysia. He has a wide academic interest in humanities and social sciences.Although he is a full-time manager, he is active in research and publishing. He has published many books and papers locally and internationally with reputable publishers and journals. He has also presented papers at local and international conferences.

Meg Milligan PhD in Sociology and Anthropology, American University of Beirut, Lebanon, PhD in Counseling Psychology, Auburn University, Alabama, USA, psychology internship at the Veterans Administration Medical Center, Tuskegee, Alabama, licensed psychologist in Alabama, is a Professor in the Department of Psychology and adjunct faculty in the global leadership doctoral program at Troy University, Alabama. Milligan has thirty years of teaching experience in higher education in Malaysia (2016-2022) and the United States, conducts study abroad programs annually, has presented over 100 papers and posters at conferences in nine countries spanning three continents, has faculty, national research, and international research excellence awards, and publications on disability rights, narcissistic leadership, and global identity among others. Milligan is President-Elect (2023-2024) of the Association for Distance Education and Independent Learning (ADEIL) after two terms on their Board of Directors, will be President (2024-2025), has Quality Matters (QM©) master reviewer certification at all levels.

Chapter 2
Edutourism in the Asia Pacific: Its Emerging Trends and Specific Features

Hoon-Peow See and Tai-Chee Wong

1 Background and Introduction

Educational tourism has its origin linked to the notion of leisure travel and learning where one experiences new cultures and interacts with peer learners and local communities. Thus, travelling to learn or to gain knowledge is not a new phenomenon. It can be traced way back in the seventh century when Japanese Buddhist monks travelled to Xi'an, China, the then prosperous capital of the Tang Dynasty where they spent years learning from Chinese Buddhist scholars, the essence of the scripts brought back by Xuan Zhuang from India. China's Tang Dynasty (AD 618–907) was also a choice destination for the Japanese and Korean intellectuals and administrators who loved to travel to China to learn the Chinese classics and administrative system. More recently in Europe during the seventeenth to nineteenth centuries, it was a common practice that young British aristocrats undertook a *"Grand Tour"* to Continental Europe with an objective to acquire new knowledge, enrich their learning experience and expose themselves to the "civilized society" (Maga & Nicolau, 2018).

Over the past five decades, however, edutourism has expanded in scale and scope as a result of rising affluence of the middle classes and the popularity in higher education, and ease of travel (Tomasi et al., 2020). This phenomenon has attracted much interest from researchers who have examined a diversity of educational activities and their particularities in different countries and regions, for example, Attaalla (2020), Indrianto et al. (2020), Mursi (2015), Ritchie et al. (2003) and Soltani (2023).

H.-P. See (✉)
Sunway TES College, Iskandar Puteri, Johor, Malaysia

T.-C. Wong
New Era University College, Kajang, Selangor, Malaysia

T.-C. Wong et al. (eds.), *Cultural Tourism in the Asia Pacific*,
https://doi.org/10.1007/978-3-031-63459-8_2

With the increasing complexity of edutourism, particularly in its role as an important business sector, defining edutourism has become more challenging and thought provoking. Bodger (1998) defined edutourism as a program whereby people move to a location with the aim of acquiring new knowledge and experience. Ritchie et al. (2003) argued that educational tourism is mainly an activity that an individual or a group of people who have made at least an overnight stay in a foreign land with learning as their primary or secondary motive.

Thus, the learning process could be anything educational in purpose varying from very formal pursuit for higher degrees lasting many years to a casual daily tour where travelers may learn briefly about local culture, history, economic activities, geography, among others (Indrianto et al., 2020). The difference in substance is hence obvious in that those trips where students are required to follow a highly structured and highly demanding program, they will have to go through a well trained process leading to a certain professional degree/diploma or a certification offer. In such a case, leisure is never a fundamental part of the study objective, whilst student life experience is highly emphasized. On the contrary, a less structured study tour of a few days up to a few weeks such as short-term language classes, summer or winter camps, topical training sessions which covers both leisure and learning does not normally offer certification of academic rigour.

The emergence of the Asia Pacific as a large global economic region, in particular, has acted as an enabling force that encourages cross-border pursuit for higher academic qualifications or knowledge acquisition and experience broadening in foreign lands. Countries or territories like Japan, China, Hong Kong, Taiwan, Singapore, and surprisingly Malaysia as a small Southeast Asian nation, have attracted tens of thousands of international students to join in their short or regular tertiary education programs in recent decades. Other countries like Indonesia have also caught up slowly in this cultural-cum-economic game.

Edutourism must be differentiated from cross-border online teaching and learning where there is virtually no physical contact between teachers and students, and basically no cultural immersion experience by students in a foreign host nation. In this conceptual context, education tourism can be defined as a process where an individual or a group of learners are engaged in a learning activity physically away from their home country. The duration of such process varies from an overnight stay to as long as multiple years for higher degree courses. To a sending country, the learners are involved in fund transfer in the form of fee payments and living expenses to the recipient country. By offering learning courses to foreign learners, the recipient country has thus generated an income from an external source which has the capacity to contribute towards its national GDP aggregate.

For a world renowned higher education institution like the University of Cambridge, its impact on the local township as well as the British economy could be tremendous. In 2020–2021, the University's economic impact through its broad range of activities was estimated at £29.8 billion, 77.5% of which was derived from the University's research and knowledge exchange activities which included business companies built out from such activities. Educational exports, associated tourism and teaching and learning activities involving international students and visitors accounted for close to £2 billion (Freeling, 2023).

This chapter explores edutourism in the Asia Pacific as a dynamic region. We first examine the emerging trends of edutourism development and the different categories of edutourism. This is followed by investigating the specific features of edutourism in selected Asia Pacific countries. Unique particularities of each country will be highlighted and analyzed.

2 Categories of Edutourism

As defined by The United Nations World Tourism Barometer (UNWTO), educational tourism is one which covers "those types of tourism which have as a primary motivation the tourist's engagement and experience in learning, self-improvement, intellectual growth and skills development. Education tourism represents a broad range of products and services related to academic studies, skill enhancement holidays, school trips, sports training, career development courses and language courses, among others" (UNWTO, 2023).

Thus, edutourism is a social infrastructure which is a rising transnational educational industry of global scale which is highly diverse in scope and learning duration. Its diversity covers a wide variety of ways of interactions in teaching and learning. The concept of learning may be in the form of informal or formal mode. Informal mode could be simply an "education with travel" which includes for instance tour packages designed more specifically for business managers who do it as a bonus-cum-leisure program paid by their respective institutions or corporations. Broadly speaking, edutourism covers the following key aspects of programs which we will elaborate from the most common to the less common or *ad hoc* ones (see Polkadotpassport, 2023; Wings, 2021).

(a) Study abroad full-time regular Tertiary level Certificate/Diploma/Degree/Graduate programs: These programs normally provide a combination of classroom learning, immersion of local cultural experiences, and field trips which provide even more exposure and understanding of program content. Students choosing to do post-graduate degrees in a foreign country would have a chance to work and learn from renowned professors and enjoy highly sophisticated facilities which their home institutions may not be able to provide.
(b) Regular full-time Primary/Secondary/Pre-university programs: These study groups are common with children of diplomats, middle-ranking or higher level executives of large multinational firms who are sent to serve in foreign countries. Pre-university programs are often enrolled by students from well-to-do families from developing countries to more developed English speaking countries, such as the United States, Canada, Australia, New Zealand, among others. These students normally spend 1-year or at the most 2 years of the pre-university program before being accepted into a local university.

(c) Student exchange programs: Student exchange programs provide students with opportunities to interact with lecturers and peer students throughout the world. Within the allowable limits set by their own institutional programs, they can earn credits towards their diplomas/degrees via the Credit Transfer System. The present practice is commonly between one to two semesters. Student exchange programs are popular among young students who are keen to see the world and mix with foreign youngsters of their own age. Nonetheless, the cost involved may be a prohibiting factor for students from poor countries who usually look for a more developed country which is costly. Tuition fees may be exempted if there is a mutual agreement for fee exemption between two exchange institutions, but the living and travelling cost could be a heavy burden for those from low-income families in the developing world.
(d) Language immersion programs: University students studying foreign languages would practically find this option an effective way to learn a language from native language speakers. Learning from a local teacher in classroom and practicing the language outside the classroom with the members of the public, nt only students would improve the spoken language they learn with a more proper accent and pronunciation, they could also change their language thinking in expression heavily influenced by their own mother tongue. As the British IELTS and the American TOEFL English are the most accepted tests for foreign language qualification, institutions offering this kind of training and teaching are very popular.
(e) Volunteer programs: Volunteer opportunities are offered by international organizations such as the World Bank and the United Nations or large companies and non-governmental organizations which provide volunteers to serve in international destinations. Offers are usually based on a short-term contract, sometimes renewable. Volunteers work in a great variety of settings from teaching in schools, nursing in hospitals to assistants or experts in certain fields useful to the local communities. During their service period, they gain a good understanding of local social and cultural issues. Youngsters taking part in these programs are able to combine travel experiences and learning, thus picking up skills and knowledge which are helpful to their future career.
(f) Cultural/Historical/Environmental tours: These tours of short-term duration are often organized with an objective of learning by travelling to the specific sites. For instance, a study group aiming to learn more about sustainability and biodiversity of a tropical natural environment may choose to travel to Brazil to visit wildlife reserves and national parks. With environmental conservation as objective in mind, travelers are highly conscious of their own ecological footprint and impact on local communities. This awareness serves not only the core theme of the learners' study trip but also their choice in having an eco-friendly accommodation and a community-based initiative that are conducive to conservation and ethical practices.

    By the same token, cultural and historical tour organizers may draw up a study plan for their participants to learn more about the history and culture of a particular destination. During the tour, they will participate in cultural activities and visit monuments, historical sites, and museums, among others.

(g) Workshops: They are organized for individuals and groups to attend workshops, symposiums, or conferences of a short stay of one to a few days where they are expected to acquire knowledge and exposure related to their field of interest. For business executives, a workshop package serves not only to help them grow professionally but also establish a network of relationship. Professional organizing bodies would normally work closely with professional institutions or government departments to create attractive programs and to invite potential participants.

(h) Internship overseas: An internship to a foreign country lasting a few months provides students before their graduation a unique insights and practical experience. As interns are paid a stipend which is usually adequate to keep them afloat during the internship, the only concern is the travel expense which is normally paid by students themselves. Internship is highly helpful to those who perform well as they are often offered a job on completion of their internship. Moreover, having established contacts and a network will also assist them in their future career, apart from broadening their worldview and sharpening their cultural awareness and intellectual growth gained from an internship done overseas.

Studies have shown that effectiveness of internship is particularly relevant to the disciplines of tourism and hospitality where interns could be trained to be highly qualified professionals with supplementary knowledge and competences that ensure their employability. The research findings by Ferreras-Garcia et al., (2020: 66) suggest that competencies that could be acquired via internship can cover generally "the application of knowledge to practice, critical and logical thought, ethics, human resources, problem solving, decision making, customer service, addressing uncertainty, oral and written communication, interpersonal communication, command of foreign languages", among others. The list may appear to be too promising and too good to be true, but it is already highly satisfactory if one could achieve just one-third of it.

3 Emerging Trends of Edutourism

Rising significance of edutourism at the global level is obvious. The same is true of the Asia Pacific as a result of rising affluence of the middle classes and the implications of higher level of education to getting high-pay jobs in the increasingly important knowledge economy. A recent study by Vuong and his partners revealed that East and Southeast Asia has witnessed that increasingly more foreign students were coming into this region and the rate of out-going students was declining Their study has indicated that incoming students were motivated by potential of higher income with high qualification, enhanced of power and political influence, prestige and reputation, enhanced knowledge in local customs and cultural values (Vuong et al.,

2021). According to another report by Future Market Insights (FMI, 2022), the educational tourism market is anticipated to reach globally a CAGR value of US$1950 billion by 2031.[1] Countries with a modernized education system, equipped with globally competitive teaching standards, smart learning programs and a well up-to-date education curricula have gained a winning edge.

In this global competition, English speaking countries with a well built up structure which is strongly supported by the integration of technology into its education system have a great advantage in attracting international students whose second language is English. Australia, despite a small population of only 25.6 million in 2021 is an exemplary case whereby it has earned a relatively large share in this new economic game.

In 2022, the revenue received from Australia's international education was estimated at A$26.6 billion,[2] of which A$2.8 billion was accountable from tuition fees from students studying online outside Australia due to Covid-19 travel restrictions. Three years earlier in 2019 before the outbreak of the Covid-19, the total revenue was about A$40 billion, about one-third higher than 2022 (Department of Education, 2023). During the Covid-19 pandemic period (2020–2022), there was also a significant impact on students' career perceptions towards higher education as well as their motivations and behaviors in selecting educational tourism destinations. Asian students' usual destinations such as Australia were affected, even though digital classrooms and transformative virtual and online learning forms were widely introduced (Scispace, 2023). As the UNWTO Barometer had shown in late 2023, global tourism had shown a remarkable resilience and sustained recovery. By the first 7 months in 2023, international tourist had achieved globally 700 million arrivals reaching 84% of pre-pandemic levels in 2019 and a 43% higher than in 2022 during the same period (UNWTO, 2023). Obviously, the trend of large intake of educational tourists including the Asia Pacific region is expected to resume soon.

Of all the programs international students undertook in January–August 2023 in Australia which numbered a total of about 860,240, higher education programs accounted for 48%, vocational and training 32%, English language courses 15%, schools 2% and other non-award programs 3%. The two key sources of sending countries were China and India, accounting for18% and 14% respectively (Department of Education, 2023).

Global travel companies in anticipating high growth in the educational tourism sector are also an important player. Though nascent in certain areas of the world, many travel companies have engaged themselves in educational tours. In this era of great socio-economic change, the concept of informal learning characterized by "education with travel" is equally appreciated as formal learning. Indeed, the thirst for knowledge supported widely by national governments and rising affluence of many social groups, has encouraged people to travel across the globe to supplement

[1] CAGR is compound annual growth rate which indicates the mean annual growth rate of an investment over 1 year or longer. The rate explains the calculation of returns in rising or falling values in individual assets, investment portfolios over a specific period of time.

[2] As of December 2023, one Australian dollar was equivalent to US$0.667.

knowledge derived from ideas and technologies different from what they had studied earlier.

Needless to say, generating revenues in response to the ever-growing world-wide demand for higher education and lifelong training has brought to the attention of the national governments by making international student recruitment an economic priority. A few of the key players in the educational tourism market are Meridean Overseas, GVI Company, Global Volunteers, Capital Tours, Inc., EF Educational Tours, Road Scholar, AAI Edutourz, ACIS Educational Tours, GoIreland, Qadri International Education Consultancy, Intelligent Partners, Futures Abroad, ProEd DMCC, Education Resources Network (ERN), Education Zone, Fact, IQ Education Consultants, M.K.H Consultancy Services, and Stratix Consultants, among others (Future Market Insights, 2022).

3.1 Future Scenario of Edutourism

Arguably, developmental trend of edutourism must be evaluated on its market potential. It is believed that edutourism has the potential for hybridization with other segments of tourism and non-tourism sectors. In this hybridization process, for an effective learning to take place, three categories of learning outcomes are esteemed to be essential. These outcomes are respectively cognitive (what to learn), affective (attitudes and ways of thinking), and behavioural (measurement of skills to be developed) (McGladerry & Lubbe, 2017).

Following this logic, future operations of edutourism are influenced to a good extent by the ways tourism education are conducted. As a worldly organization, UNWTO's contemplative view in the trend of tourism education should be seen as most worthy of consideration. UNWTO (2022) has suggested seven areas of tourism education that are believed to have a "sustainable, innovative, and inclusive" quality to move towards a meaningful provision of quality tourism education. These seven areas served as guidelines were aimed to guide UN member states in their national policies in creating a tourism educational pathway to train up their vocational and managerial career-takers in tourism, as follows:

(a) Increase access to tourism education: To offer courses, webinars and digital resources by different training units on different topics, based on recommendations by UNWTO or partner institutions and experts;
(b) Update tourism curricula: Updating is necessary as to scale up courses to respond to the rapidly changing tourism sector due to digititalization and global pandemics for example;
(c) Integrate tourism education into secondary school curricula: To incorporate vocational tourism education and practical tourism into existing secondary school systems. Training, internships, and tourism educational materials used for teaching will help train tourism workers into the workforce;
(d) Innovate undergraduate and graduate tourism education: To introduce new majors such as Tourism Data Science, Tourism Education Policy, International

Relations and Tourism Studies, Public Policy and Tourism, and Tourism Business Management;

(e) Standardize certifications: This standardization facilitates easy recognition of education standards globally for employment purposes in the area of tourism knowledge;

(f) Expand tourism education quality assurance: This serves the primary objective of quality control of tourism education standards; and

(g) Connect tourism graduates with employment opportunities: Quality and standards of tourism graduates are yardsticks towards employment provided by corporations and governments to help in their global business competition, and hence enhance their service quality in the tourism sector.

One of the UNWTO's (2023) sustainable development goals is to overcome challenges facing tourism as a large employment sector with long-term objective in ensuring quality job provision through the transformation of the tourism education ecosystem. In this aspect, apart from international collaborations, public-private partnerships within nations are also critical in creating a resilient tourism sector with the help of educational institutes. With such a plan, there will be a high success rate in the provision of quality services in tourism training inclusive of educational tourism with the support of technology.

Like practically all academic disciplines that involve teaching, learning, research and a job market, edutourism is no exception in its developmental path that will bring change along the way. The key elements that will influence its pathway are those associated with its relevance to the social and economic need of human society and in an evolution of knowledge building in an ever-evolving world. The shift is expected to undergo a transformational shift in the short- and medium-term, and probably a paradigm shift in the long term.

Market Research Future (MRF, 2023) has provided a visionary forecast in the future trend of edutourism. Whilst edutourism is expected to continue sustainably and responsibly from its current pattern inclusive of those such as studying in foreign countries where the programs are relatively advanced and up-to-date, cultural immersion programs which enable learners to engage with local community members, acquiring a deeper level of foreign languages and better understanding of different cultures. For the latter, home-stays, language exchange programs, and visits to indigenous communities are exemplary for authentic cultural interactions that enrich learners' experience. Other features that MRF has highlighted are outlined as follows.

(a) Digital integration: This technology is already in use but it will be further refined and made more sophisticated and user-friendly to reshape and support the educational tourism teaching and learning. Virtual tours, historical reconstructions, and interactive lessons will be increasingly designed to provide visual and mental impact to learners prior to or during their visit to the destination site. These tools will serve as an innovative bridge to fill in the gap between classroom learning and the real-world experience.

(b) Personalization: Where resources allow, personalized educational trips are expected to be made increasingly available to group learners who can afford to pay more for professional training services. Learners of this category can be offered study trips and programs based on their specific interests, preferences, and learning goals. Consequently, customized travel plans matching individual educational objectives will be strengthened to ensure a more meaningful learning experience.
(c) Lifelong learning: Educational tourism is expected to be extended to more people rather than students or young travellers. In view of the fast-changing scope and knowledge content of many tech-related disciplines in the job market, professionals, managers, and employees will have to cope by expanding their knowledge and skills throughout their lifelong career. New knowledge is particularly important for mid-career shifters to acquire through training in the form of workshops that meet their specific needs.
(d) Cross-disciplinary learning: It is commonly known that the boundaries between different disciplines of study have become increasingly blurred these days in the job market. As such, interdisciplinary learning experiences that combine different subjects, such as art and science, history and technology, or ecology and anthropology have become more and more important. The cross-disciplinary approach offers to train professional or managers to equip themselves with a more comprehensive understanding of complex issues.
(e) Global collaborations: Knowledge know no frontiers. In international business collaborations, we see a rising trend that corporations, academic institutions, educators, and learners undertake exchange programs, joint workshops, and collaborative research projects to learn from each other. Such collaborations promote global cultural exchange and the advancement of knowledge at a global scale.

As we move on from here, learning beyond borders will be an integrative force that helps in bringing cultures of different origins and languages closer to each other. With the help of more affordable means of international transport, meeting up and interacting with other nationals in foreign soils will be made easy. More particularly, such interactive teaching and learning meetings, whether at home or abroad will be poised by technology as an enabling device. This includes virtual reality (VR) and augmented reality (AR) methods which will enable learners to immerse themselves in the educational events, and interact with distant cultures in a comfortable environment.

4 Specific Features of Edutourism in Selected Asia Pacific Countries

English being the most globally used language in business, science and technology and the academia has the strongest appeal to international students. In this domain, the United States leads the world in attracting international students and educational tourists, followed by the United Kingdom and Australia. The United States attracts large numbers of international students in both formal tertiary qualification programs and short-term edutourism programs such as student exchange, professional and scholar exchange, short-term certification and executive development and internship training. Other developed countries, Germany and France in Europe and Japan in Asia, are the closest contenders, despite the fact that their language represents a barrier to many (Attaalla, 2020: 35; Education USA, 2023).

In the Asia Pacific region, other than the developed nation Japan, a number of other Asian countries are catching up very quickly in attracting international students. We shall consider a few of them and analyse their achievements below, with highlights placed on structured short-term programs.

4.1 Japan

Japan is a popular destination for international students in Asia. In 2015 alone, Japan attracted 65,000 students, from Taiwan, Australia, China, Korea, United States and others (Attaalla, 2020: 37). Way back in 2008, Japan aspired to increase its international students to a total of 300,000 by year 2020 (Student Exchange Office, 2008), and it achieved by then 312,000 international students exceeding its aspired figure. Without exception, Japan had to face a sharp fall in number during 2020–2022 due to the Covid-19 pandemic (Statista.com, 2023).

In 2008, among the international students in Japan, Chinese students outnumbered the total from the rest of the world. This is not surprising given the size of China's population, and its proximity to Japan. Historically, during the late nineteenth and early twentieth century, reformist Japan undertaking modernism was a role model to many Chinese students who dreamed to remake their weakened motherland a strong nation again. Quite a number of Chinese prominent figures such as Sun Yat-Sen, the founder of Guomindang and Lu Xun, a famous literary man did study in Japan (Harrell, 1970). Among the 118,498 international students studying in Japanese universities and other institutions in 2008, only 8368 (7.1%) were short-term students registered mainly in exchange programs (Student Exchange Office, 2008: 10, 36).

Other than exchange programs, Japan's summer short courses in subjects such as sustainable urban management, international relations, nanoscience are highly popular to foreign students. For example, Keio University offered the following summer courses in 2023 (Keio International Center, 2023) as follows:

(a) Business and Economy: Entrepreneurship and Small Business Development in Japan;
(b) Culture and Cross-cultural Understanding: Courses include "Understanding Japan through cross-cultural comparison", "Otaku Culture in Japan and Its Transnational Relation", "Seeing Japanese Culture and Society from a Qualitative Perspective" and "Culture and Emotion";

(c) Language and Literature: "Ghosts and Goblins in Modern Japanese Literature";
(d) Arts: "Introduction to Japanese Art History", "Arts-Art Workshop- Discovering Arts and Culture in Japan";
(e) Law: "Japan's Trade Law and Policy", "The Supreme Court of Japan- Structure, Function, Case Law";
(f) Religion & Thought: "Japanese Buddhism and Social Suffering", "Sociology of Education"; and
(g) Japanese Language: "Business Japanese- Introduction to Japanese for Business and Work", "Survival Japanese- Introduction to the Japanese Language".

Although Japan's short-term courses only manage to attract a relatively small number of international students, its most developed and most successful economy status in Asia has remained attractive to many in the developing world. It has stayed on as a force to be reckoned with in edutourism backed by its longstanding unique culture and a much organized tertiary education structure.

In recent years, Japan has to face competition from the fast developing China, putting her perhaps somewhat behind the limelight, especially in the area of high speed rail infrastructure. China has proved to Japan not only in its superiority in speed but also in its extensive coverage in a much larger national territory.

4.2 *China*

Prior to China's opening up from the 1980s, Taiwan was a traditionally popular destination in higher education for mandarin speaking students from Southeast Asia. However, in the past 20 years China has gradually overtaken Taiwan as a provider for Chinese language-based higher education. By 2018, China has become one of the most popular destinations for international students, recording 492,185 international students enrolled in its higher education institutions, overtaking Japan in Asia (Erudera, 2023). One has to be reminded that China is also the biggest supplier of international students in the world.

As a rising power, China has begun over the past 30 years to offer scholarships or financial aids to international students, particularly to those from Southeast Asia who have the capability to pursue diploma and degree programs conducted in Chinese language. According to Billman's (2022) estimate, about half of the international students in China are in short-term programs, notably in short Chinese language courses accounting for about half of all international students. During the pre-2020 epidemic period, young foreign and moderately wealthy students were heavily attracted by an "experience-seeking" venture, looking after an exciting cultural exposure. In response, Chinese universities have offered unique and flexible programs catering to their need.

Among the short-term programs, there is a very popular one called simply "Doing Business in China". As the "world factory", China plays a prominent role in international trade. Business representatives and executives from around the world need to have a good understanding how to conduct business in China which is vital to their corporate success or failure. The courses are taught in English by both faculties from London School of Economics and Peking University provides participants with the critical tools inclusive of subject-specific modules, company visits, and cultural activities. By merging classroom knowledge and the real world practices, students will benefit from China's business success experience in its rapid economic growth. In this matter, the program of Peking University's (2019) Guanghua School of Management (2023) states this:

> the 2-week program provides you with a unique opportunity to experience firsthand the emerging markets of China, one of the world's fastest growing major economies, offers a unique angle on China for those well-versed with China as well as those new to China. The Final Project best reflects the culmination in-class learning and company visits. This project is a highly collaborative exercise that gives participants the opportunity to apply the concepts and techniques learned through the program to a unique, self-selected China business initiative, weaving your learning across the program into your practice.

Apart from offering Chinese language courses to foreign students at home, China also followed the steps of the United Kingdom and France by setting up language training centres overseas. Starting in 2004, China launched its first Confucius Institute in Seoul, South Korea. By 2019, there were a total of 530 Confucius Institutes worldwide, and they operate largely in collaboration with local universities in teaching Mandarin Chinese and Chinese culture. Despite being caught in controversies, especially in the West, the Confucius Institutes' achievement in promoting China to international students cannot be ignored (Confucius Institute, 2020).

4.3 Malaysia

Despite being a relatively small and less developed economy, Malaysia is unique and has managed to attract a large number of edutourists. Various media reports put the total international student number at 130,000–170,000 for 2022. Malaysia has a long-term goal to build that enrolment base to 250,000 students by 2025 (Icef Monitor, 2023). According to Malaysia's Ministry of Higher Education, as of May 2022, there were as many as 585 public and private universities/colleges in Malaysia, of which 73% were private. Though about 330 of the private institutes belong to small training centres or small colleges, the rest are formally recognized universities and university colleges offering diploma to degree to PhD programs (Wong, 2022). They are popular with international students given that virtually all programs are conducted in English language and their fees are relatively cheap compared to developed Anglophone countries.

In 2015, The Ministry of Higher Education Malaysia in collaboration with Tourism Malaysia officially launched "Malaysia 101 Edu-tourism Package (M1EP)" to boost the edu-tourism industry. Accordingly, the Ministry's objective was to establish a strategic partnership that "enables the Public Higher Education

Institutions to assemble and introduce educational tour packages to the public and the international community so as to provide a more holistic education, in line with the concept "Beyond a Degree", that will enable them to plan trips in the country" (Mursi, 2015). Foreign students pursuing a program of less than 3 months would be issued a social visit pass, instead of a harder to get student pass, said the Ministry's spokesman.

Responding to the appeal by the Ministry of Higher Education, Universiti Putra Malaysia was the first university to organize and offer short-term education packages to international students. In his observation, Lee (2021: 3) revealed that these programs on offer were simple packages involving normally some form of English learning, day trips to popular tourist sites, a short introduction of Malaysia and Malaysian culture, and where internship is involved which lasts normally a longer period, industry visits.

Such short programs are sometimes called summer programs as they are often offered during summer in the northern hemisphere. On other occasions they can be known simply as edutourism programs which may last for 1 year long. For example, the International Relations Centre (UKM Global) of University Kebangsaan Malaysia (UKM) or National University of Malaysia has designed an International Summer Program offering short-term courses and cultural immersion lasting from 1-week to 1-month. Other short programs offered by UKM are given below (International Relations Centre, 2023).

One Week Program: Intensive Islamic Finance, Future Global Leadership;

Two Weeks Program: Management, Leadership and Culture Immersion Program, Economics and Business Management Practices in Malaysia, Tropic Nature and Heritage, Field Entomology and Parasitology for Beginners, Crime Scene Investigation

Three Weeks Program: Indigenous Community of Malaysia, Sustainable Tropical Heritage, Islamic Civilization: Malaysia Perspectives, The Power of Future—Fuel Cell

Four Weeks Program: Breakthrough Intensive English Language, Engineering and Sustainability

Other private universities and foreign universities branch campus offer similar but more limited number of programs, depending on their scope of investment and academic strength. These programs allow opportunities for students to learn conversational English. This is especially popular among hospitality students and practitioners. From personal observation, these programs as a whole lack depth from the cultural perspective, if compared to the "Doing Business in China" program and other summer programs found in Japan and China. Malaysia has neither the advantage of teaching technology. More often than not, Malaysia is chosen by international students for these short programs mainly in consideration of its lower cost.

4.4 Indonesia

As a large and populous country with 250 million people, Indonesia in 2016 had domestically 4,941,570 higher education students. But the country had merely 6000 international students, if compared with much smaller countries like Malaysia and Singapore which had respectively over 60,000 and 50,000 in the same year. The factors behind such a low figure could be attributable to the relatively poor quality of Indonesian educational institutions, lack of student exchange programs, official publicity and student visa issues (Global Business Guide, 2016). Language is most likely an additional factor as Indonesian is hardly learned by students across the world outside Indonesia and Malaysia. This scenario is likely to change in the not too distant future if Indonesia climbs up the technological ladder and becomes an increasingly important global economic player.

To overcome the weaknesses in education quality of Indonesian local universities, the Indonesian President Joko Widodo has called for the improvement of the Indonesian academia and higher education performance. In facing this tough challenge within its higher education sector, Indonesia has placed strong emphasis on putting substantial resources to develop the country's human capital through education and health improvement. For Indonesia, one of the approaches is to offer English as the language of instruction for international students. For example, the Bandung Institute of Technology, Binus University and Universitas Prasetiya Mulya have begun to offer undergraduate as well as MBA programs in English to international students. With an expanding economy, Indonesia's rich cultural heritage and language represents an attractive attribute to international investors including the multinationals seeking to penetrate the Indonesian market, and those who are interested in Indonesian cultural inquiries (Jakarta Post, 2023; Global Business Guide, 2016).

As part of the cultural exploratory effort in edutourism development, University of Surabaya has designed a community-based tourism strategy which is made up of three aspects of activities covering eco-tourism, heritage tourism and rural tourism. In the study tour, participants are guided in a rural setting with exposure to an "architectural trail, bio-diversity, animal watching and happy farming" agenda. During the early stage of training, local tour guides would serve as instructors in providing knowledge about the local community's daily practices (Megawati et al., 2016: 1–2).

4.5 Singapore

Since its independence in 1965, the city-state has adopted a pro-Western and export-oriented industrial policy heavily dependent on global trade in which English has a strong advantage in business dealings and technology transfer. Having adopted an elitist and pragmatic style of governance, the government led by the People Action

Party has placed heavy investment in education and international talent recruitment which has brought in reward in making her a high world ranking leader in higher education over the last two decades.

Known as one of the global education leaders for its quality, Singapore in 2022 had about 55,000 foreign students from both the developed and developing world. They came predominantly from Malaysia, Indonesia, Thailand, Vietnam, China, India, and South Korea. At the same time, there were around 21,660 Singapore students studying abroad, 18% of them studied in the United States. Beside offering places to foreign students, Singapore has also attracted a considerable number of leading international institutions to set up joint degree or exchange and research collaboration programs with local universities and colleges (ITA, 2023). This in turn has helped enhance the city-state's attractiveness, thus creating a strong demand from students in the region to study there.

In the area of short programs, there are more professional training services as per state policy to upkeep the relevance of its workforce than attracting foreigners with profit-led programs. For this purpose, MNCs and strong government-linked companies in Singapore often organize short courses, like leadership, project management, sigma quality, digital transformation, and other continuous improvement programs to train their staff on the job. Small and medium enterprises are eligible to government grants to cover the costs when they intend to have their staff undergo such courses (ITA, 2023). A very peculiar program which is called the "Mayors' Program" is worthy of discussion as it is conducted specifically for Chinese senior pubic officials.

4.5.1 The "Mayors' Program"

After China's paramount reformist leader Deng Xiaoping visited the United States in early 1979, he realized there was an urgent need to send students to the developed West to learn up-to-date managerial and technological skills. The trend shifted to Singapore in the early 1990s in managerial domain based on two factors that Singapore shared with China. First, Singapore had quite an identical Confucian tradition like China, and they both had a one-party rule with values more of a top-down command and control approach. Second, courses could be conducted in both English and Chinese effectively, and mandarin Chinese was particularly welcomed by Chinese senior officials as they were not well versed with the English language used in the West.

From 2001, "Mayors' Classes" designed for Chinese senior officials started in Singapore's two leading universities, National University of Singapore's Lee Kuan Yew School of Public Policies and Nanyang Technological University. Primarily, the program aims to raise the administrative efficiency of Chinese public administrators, and to comprehend how the Singapore style of governance was able to run state-owned enterprises and public services efficiently with high pay to achieve corruption free (Qilu Wanbao, 2011).

From the Nanyang Technological University records (NTU, 2023), we learn that the first Executive Training Program for Chinese officials was introduced in 1992. Batches of officials in groups of 10–20 subsequently came and went, and courses would last from a couple of weeks to a few months, depending on the depth and rigour of each course. While the scope of the program has declined in intensity over the years, it has lasted until today.

There are also longer study programs. More on the basis of inter-governmental collaboration, Singapore Ministry of Trade and Industry initiated in 1998, the Master of Science in Managerial Economics, and in 2005, the Ministry of Education launched the Master of Public Administration and in 2009, the Ministry of Manpower came up with another Centre for Public Administration (NCPA). Together with a few other Master's programs set up till 2021 taught by the NTU faculty, they welcome senior administrators from China and other Southeast Asian nations to attend. By 2022, the total of full-time students from overseas attending NTU's Master's programs cited above accounted for 500 (NTU, 2023).

5 Conclusion

The discussion in this chapter indicates a strong positive correlation between economic development and edutourism and this trend is anticipated to persist in future in a more refined manner. In the Asia Pacific region, having capitalized on popularized air transport and education expansion, edutourism has witnessed a sharp rise in activities as an attractive destination whereby it has played the role of an "economic miracle" contributing significantly to the regional GDP growth.

Japan being the first developed Asian country has traditionally been attractive to international students and cultural tourists alike. This has gradually shifted to China after she has executed reformist measures over the past 40 years which have brought in a high GDP growth miracle and helped achieve a higher academic standard. Japan and China are not the only countries that are attractive in edutourism, a number of smaller Asian nations, particularly Malaysia and Singapore too have attracted a large number of international students in both regular and short course programs, thanks to their prompt adaptation capacity to meet prevailing demands and standards. Another country like Indonesia, with a large population, territory and rich culture should also appeal to foreign students in the near future despite its current weaknesses in standards and lack of facilities.

In the prospect that advanced and virtual educational technology will provide tremendous help in upgrading edutourism whether in a chosen foreign destination or via across-border online teaching, maintaining a robust economic development pattern alone is certainly not enough to attract and retain edutourists in the long run. There is an urgent need that greater effort be rendered in creating quality regular and short-structured programs to attract young international students and business executives/tourists for in-service training and skills upgrading. Cross-border edutourism, after all, will play a positive role in narrowing down the educational gaps as well as bringing closer cultural ties between nations.

References

Attaalla, F. A. H. (2020). Educational tourism as a tool to increase the competitiveness of education in Egypt: A critical study. *Journal of Travel, Tourism and Recreation, 2*(I2), 34–41.

Billman, S. (2022). How the pandemic transformed China's international applicants. China admissions. Dated 7 July 2022. https://www.china-admissions.com/blog/pandemic-china-international-students/

Bodger, D. (1998). Leisure, learning, and travel. *Journal of physical education recreation & dance, 69*(4), 28–31. https://doi.org/10.1080/07303084.1998.10605532

Confucius Institute. 2020. *Why cooperate with us?* https://ci.cn/en/hzhb

Department of Education. (2023). *Education export income, calendar year 2022*. Australian government. https://www.education.gov.au/international-education-data-and-research/education-export-income-2021-calendar-year

Education USA. (2023). *Your 5 steps to U.S. studies*. https://educationusa.state.gov/your-5-steps-us-study/research-your-options/short-term/short-term-programs

Erudera. (2023). *China international student statistics*. https://erudera.com/statistics/china/china-international-student-statistics/

Ferreras-Garcia, R., Sales-Zaguirre, J., & Serradell-Lopez, E. (2020). Competences in higher education tourism internships. *Education + Training, 62*(1), 64–80.

FMI (Future Market Insights). (2022). Educational tourism market with current trends analysis and forecasts 2031, dated April 22. https://www.digitaljournal.com/pr/educational-tourism-market-with-current-trends-analysis-and-forecasts-2031

Freeling, A. (2023). *Cambridge University's economic impact*. Dated March 20. https://www.cam.ac.uk/stories/cambridge-economic-impact

Global Business Guide. (2016). *Studying abroad in Indonesia for international students*. http://www.gbgindonesia.com/en/education/article/2016/studying_abroad_in_indonesia_for_international_students_11702.php

Guanghua School of Management. (2023). *Doing business in China*. Peking University. https://pkudbic.com/wordpress/wp-content/uploads/2019/09/PKU-Guanghua-Doing-Business-in-China-Program-2020-HD.pdf

Harrell, P. S. (1970). *The years of the young radicals: Chinese students in Japan, 1900–1905*. Dissertation for doctor of philosophy, Department of Philosophy, Columbia University https://www.proquest.com/openview/0e12f098dc0703113028807c485adc80/1?pq-origsite=gscholar&cbl=18750&diss=y

Icef Monitor. (2023). Malaysia exceeds target for new international student applications in 2022. 1st Feb 2023. https://monitor.icef.com/2023/02/malaysia-exceeds-target-for-new-international-student-applications-in-2022/#:~:text=While%20neither%20EMGS%20nor%20the,and%20170%2C000%20students%20for%202022

Indrianto, A. T. L., Mitasari, R., & Hangga, F. R. (2020). Educational tourism: What strategies needed to develop it? (a case study of Baobab Safari Resort). *Advances in Economics, Business and Management Research, 175*, 11–14.

International Relations Centre. (2023). *UKM international summer programme*. https://www.ukm.my/pha/summer/

ITA (International Trade Administration). (2023). *Singapore education and training services industry snapshot*. https://www.trade.gov/country-commercial-guides/singapore-education-and-training-services-industry-snapshot

Jakarta Post. (2023). *Road less traveled: Indonesia through the lens of international students*. Dated February 14. https://www.thejakartapost.com/culture/2023/02/14/road-less-traveled-indonesia-through-the-lens-of-international-students.html

Keio International Center. (2023). *Keio summer program 2023: Overview*. https://www.ic.keio.ac.jp/en/study/short/summer_program/summer_program_information.html

Lee CK (2021) *Edu-tourism: Leisure motivation, travel behaviour and travel satisfaction among foreign students in Malaysia*. A thesis submitted to Asia e University, in fulfilment of the requirements for the Degree of Doctor of Philosophy.

Maga, A. A., & Nicolau, P. E. (2018). Conceptualizing educational tourism and the educational tourism potential (evidence from ASEAN countries). *Advances in Economics, Business and Management Research, 39*, 343–348.

Market Research Future (MRF). (2023). *The future of educational tourism: Where travel and learning converge*. https://www.linkedin.com/pulse/future-educational-tourism-where-travel-learning

McGladerry, C. A., & Lubbe, B. A. (2017). Rethinking educational tourism: Proposing a new model and future directions. *Tourism Review, 72*(3), 319–329. https://www.emerald.com/insight/content/doi/10.1108/TR-03-2017-0055/full/html

Megawati, V., Rejeki, E. S., Gunawan, Simanjuntak, Y. N., & Krisdinanto, N. (2016). The development of educational tourism in the area of Mount Penanggungan through community empowerment. In *13th Ubaya international annual symposium on management* (pp. 1–12).

Mursi, M. (2015). Launch of Malaysia 101 edutourism packages. *GayaTravel Magazine*. August 15. https://www.gayatravel.com.my/launch-of-malaysia-101-edutourism-packages

NTU (Nanyang Technological University). (2023). *Our milestones*. https://www.ntu.edu.sg/ncpa/about-us/milestones

Peking University. (2019). *Doing business in China*. Guanghua School of Management. https://pkudbic.com/wordpress/wp-content/uploads/2019/09/PKU-Guanghua-Doing-Business-in-China-Program-2020-HD.pdf

Polkadotpassport. (2023). *What is educational tourism? (8 examples)* https://polkadotpassport.com/educational-tourism/

Qilu Wanbao. (2011). The "mayor program": A learning experience in Singapore. Dated May 19. https://epaper.qlwb.com.cn/images/2011-05/19/A17/qa1719.pdf

Ritchie, B., Carr, N., & Cooper, C. (2003). *Managing educational tourism*. Channel View Publications.

Scispace. (2023). *How Covid-19 affects educational tourism?* https://typeset.io/questions/how-covid-19-affects-educational-tourism-4ivo81pq1b

Soltani. E. (2023). Where do international students in the U.S. come from? *Visua Capitalist*. https://www.visualcapitalist.com/cp/international-students-in-the-u-s-come-from/#:~:text=The%20International%20Student%20Population&text=After%20a%20drop%20in%20recent,colleges%20and%20universities%20last%20year

Statista.com. (2023). *Number of international students in Japan from 2013 to 2022 (in 1,000s)*. https://www.statista.com/statistics/1125812/japan-total-number-foreign-students/#:~:text=The%20total%20number%20of%20international,declined%20for%20three%20consecutive%20years

Student Exchange Office. (2008). *Outline of the student exchange system: Study in Japan and abroad*. Student Exchange Office, Student Services Division, Higher Education Bureau, Ministry of Education, Culture, Sports, Science and Technology, Japan. https://www.mext.go.jp/a_menu/koutou/ryugaku/081210/001.pdf

Tomasi, S., Paviotti, G., & Cavicchi, A. (2020). Educational tourism and local development: The role of universities. *Sustainability, 12*, 6766. https://www.mdpi.com/2071-1050/12/17/6766

UNWTO. (2022). *Tourism education guidelines*. https://doi.org/10.18111/9789284423941

UNWTO. (2023). *International tourism swiftly overcoming pandemic downturn*. Dated 19 Sep 2023. https://www.unwto.org/news/international-tourism-swiftly-overcoming-pandemic-downturn

Vuong, Q.-H., Pham, H.-H., Dong, T.-K.-T., Ho, M.-T., & Ding, V.-H. (2021). Current trends and realities of international students in east and Southeast Asia: The case of China, Hong Kong, Taiwan and Malaysia. *International Journal of Education and Practice, 9*(3), 532–549.

Wings. (2021). *Education tourism: Courses, colleges, benefits, examples*. https://leverageedu.com/learn/education-tourism-courses/

Wong, T. C. (2022). Address rapid expansion of higher education. *The Sun*, 22 November issue. https://thesun.my/opinion_news/address-rapid-expansion-of-higher-education-HG10253532

Hoon-Peow See holds a PhD in Sociology and Anthropology from the International Islamic University Malaysia, a Master of Philosophy in Sociology from the University of Malaya, and an LLB (Hons) from the University of Glamorgan (now the University of South Wales). He has held various leadership positions in the education industry with exposure in both academic and business operations, which include being CEO cum Vice Chancellor of a university college. He is currently the Director at Sunway TES College Johor Bahru, Malaysia. He has a wide academic interest in humanities and social sciences. Although he is a full-time manager, he is active in research and publishing. He has published many books and papers locally and internationally with reputable publishers and journals. He has also presented papers at local and international conferences.

Tai-Chee Wong has his BA and MA degrees in Urban and Regional Planning from the University of Paris and earned his PhD in Human Geography from the Australian National University. After teaching 20 years at Nanyang Technological University, Singapore, he retired in 2013. He then worked as Distinguished Professor for two years at Guizhou University of Finance and Economics, China, and as Dean and Professor at the Southern University College, Johor Malaysia, from March 2017 to December 2018, and as Deputy Vice-Chancellor at Berjaya University College, Kuala Lumpur from October 2022 to September 2023. He was a Visiting Professor at the University of Paris (Sorbonne IV) and a Visiting Fellow at Peking University and Tokyo University. His main research interests are in urban and economic issues and, more recently, in Malaysian politics. Besides his 16 self-authored and edited book volumes, he has written over 100 academic articles and published widely in international journals including tourism papers.

Chapter 3
Tourism Development of Singapore: Artificiality Versus Profitability

Tai-Chee Wong

1 Introduction

Urban tourism is integral of urban cultural activities including heritage and it plays an increasingly significant role in advanced capitalism where it promotes new patterns of consumption in the interests of profit (Fox, 2023). Within the city, there are two key components of heritage, namely natural heritage and cultural heritage. Natural heritage includes features such as mountains, forests, rivers, flora and fauna species whilst cultural heritage covers man-made structures or landscapes that are designed to serve specific purposes including tourism attractions. These two categories of heritage are in short supply in Singapore, a small island state of only 700 square kilometres which has a relatively short history and a nature area much of which has to give way to rapid urbanization and infrastructure development over the past 60 years.

The developmental state of Singapore has followed an economic priority goal since its independence in 1965. Since then, it moved on successfully from its initial labour intensive and export-driven industrialization, to capital-intensive industrialization from the 1980s, and from the 2000s, towards knowledge-intensive activities (Wong & Yap, 2004). In the service sector, it had moved away from its early primarily low-ordered services to a higher-value added financial and professional services. Tourism, as an important component of the services, is equally a target pillar where much emphasis and commitment has been devoted and made into a vibrant and dynamic area of economic development.

Created in 1964 as a statutory board under the jurisdiction of the Ministry of Trade and Industry of Singapore, the Singapore Tourism Board (STB) started

T.-C. Wong (✉)
New Era University College, Kajang, Selangor, Malaysia

T.-C. Wong et al. (eds.), *Cultural Tourism in the Asia Pacific*,
https://doi.org/10.1007/978-3-031-63459-8_3

Table 3.1 International tourist arrivals and contribution as share of GDP in Singapore, 2011–2020

Year	International tourist arrivals[a] (millions)	Tourist direct contribution as share of GDP[b] (%)
2011	13.17	4.15
2012	14.50	4.19
2013	15.57	4.08
2014	15.10	4.16
2015	15.23	3.89
2016	16.40	4.17
2017	17.42	4.12
2018	18.51	3.97
2019	19.11	4.00
2020	2.74	4.04

Source
[a]United Nations (2022)
[b]Singapore Department of Statistics (2022)

humbly with a small intake of tourists. By 2011 till 2019, as shown in Table 3.1, the city-state's tourism sector had contributed around 4% to its gross domestic product and attracted during this period from 13.17 million to 19.11 million international tourists (United Nations, 2022; Singapore Department of Statistics, 2022). Until the international arrivals were badly affected by the Covid-19 pandemic, tourism had played an essential role in making Singapore as a vibrant global city with a wide variety of tangible and intangible leisure options and customized services.

Singapore's tourism has transformed itself dramatically with rapid urbanization taking place over the last 50 years. As a result of this transformation, it has turned to urban tourism which is either heritage-based or man-made leisure-related choices such as theme parks, casinos, Formula-one car racing and business tourism such as conventional conferences/events and integrated resorts. Such move is known as quality tourism with a corporate identity that places emphasis on innovative capacity and international engagements with greater potential to generate stronger economic impulses. This approach of quality tourism signifies equally the collaboration between public agencies and the private sector in enhancing competitiveness, with STB offering financial support to private enterprises various grant schemes and tax incentives (Singapore Tourism Board, 2020).

This study intends to explore the nature of urban tourism and the ways it is developed to reinforce its capacity to generate revenues and jobs. We then examine the types of urban tourism, heritage-based or artificial, which have been put in practice and won Singapore an international place. Despite success stories, we eventually dig into the weaknesses of cultural tourism which are "man-made" in origin, citing casino as an example.

2 Urban Tourism and Its Nature

The nature of urban tourism implies tourism as a whole is largely urban-based as it encompasses not only business or study-related travelers but also numerous urban destinations. The destinations can be provided with touristic images that have "a set of visual symbols and descriptors that provide visitors and residents with a transparent and recognizable local iconography for interpreting the cultural attractions of a city (Gotham, 2007: 308). Urban destinations such as tourist-historic urban cores, museums, city waterfronts, theme parks and specialized precincts have potential to attract local and international tourists in for consumption activities (Edwards et al., 2008; Murillo et al., 2013). Such destinations, like any tangible products, are however subject to the quality of maintenance, and from time to time, the consistent need for re-imaging to stay relevant in response to market demand.

Making destinations relevant to changing market demand is a challenging task. A supply-side approach suggested by Pearce (2001) had involved a set of issues that have influenced urban tourism activities such as demand, supply, marketing, planning, organization, impact assessment and operations. With regards to the nature of supply, Pearce (2001) argued that this would vary in different scale levels, which are very much dependent on policy making, management and other practical operations and applications.

Given the large scope of concerns over urban tourism, this chapter would choose to discuss first the sustainability issues in relation to marketing activities. It would be intriguing to examine whether marketing is sustainable in business operations or in need of adjusting to tourist tastes which might change over time. This is especially pertinent when Singapore's tourism is examined as a commercial undertaking where it involves primarily "a range of products to sell, market to identify and access, investment to be undertaken, product distribution to be managed, and profits to be made" (see Edwards et al., 2008: 1037). After looking at the sustainability of tourism in this pragmatic setting of Singapore, we pursue to study the reimaging and place marketing efforts in the city-state on the basis of both heritage and artificial tourism sources. The weaknesses of artificial tourism are finally dealt with.

2.1 Urban Tourism Sustainability Issue

According to the United Nations World Tourism Organization (UNWTO), urban tourism is about tourism activities taking place in an urban space with non-agriculture-based economic attributes. It possesses nodal points of transport, and destinations offering a broad and heterogeneous range of cultural, architectural, technological, social and natural products for leisure and business. Leisure making, as studies show, is healthy exercise as it helps relieve pressure, especially for those working in large congested cities (UNWTO, 2019a; Lehto & Lehto, 2019). Rising concentrations of city population which is expected to reach 60% in the world by

2030 indicates further urban tourism will be a great driving force that has an increasing impact on the socio-economic and cultural life of urbanites and future urban development policies.

One of the key goals set out by the New Urban Agenda in the 8th UNWTO Global Summit on Urban Tourism in 2019 was its Goal 11 which aims to make cities and human settlements "inclusive, safe, resilient and sustainable" (UNWTO, 2019b). Our interest in this study is about how tourism should be intrinsically linked to city activities and developed in a sustainable manner, thus providing better living conditions to its residents and visitors.

More interestingly in the context of Singapore, an advanced global city moving fast in the applications of high tech, we look at its development path towards "Smart Cities, Smart Destinations", a theme set by UNWTO in the 8th UNWTO Global Summit on Urban Tourism in Kazakhstan on 9–12 October 2019. The Summit gathered representatives from over 80 countries to address complex urban tourism challenges, the most critical of which was sustainability. The medium to long-term vision of the Summit is to enable urban tourism to be advanced into "a true contributor to the development of inclusive, resilient and sustainable cities" (UNWTO, 2019b).

According to Cardenas et al. (2015), sustainable tourism covers six principles, namely resource preservation, environmental education, stakeholder inclusion, economic planning, cultural awareness and community response identification. It is therefore a very broad-based concern as they encompass from macroscopic care of resource use, economic planning, educating people to arouse their environmental awareness, to microscopic inclusivity of ordinary people as stakeholders in the tourism industry where they share a common interest with each other.

What does sustainable tourism measure in the city context then? From the eco-centric perspective, the measurement is placed more on the "quality of life" rather than the "standard of living" whereby the measurement of economic growth has relatively little value. The ecological view is represented by concept of deep ecology which favours, among others, local communities and localized decision-making, use of appropriate technology, and it recognizes the "right of nature" independent of humans in city tourism activities.

Another element highly valuable in building sustainable tourism in cities is their resilience. By resilience, we mean the capacity of city residents-cum-visitors, institutions and systems in operation within the city how they could adapt, survive and thrive in situations of chronic stress and acute shocks (see Donelli et al., 2021: 177). In this sense, we have to inquire what actually is the cultural value in the urban fabric in which we can build up resilience. Resilience symbolizes the ability to adapt to new socio-economic circumstances when a city changes its fabric outlook from an industrial to a service-led city, for example. With the rising significance of culture-based tourism these days, this resilient adaptability implies the importance of promoting cultural heritage and enhancing its value while wealth is being created in terms of growth in business, employment and tax expansion for the governments at different levels.

Where culture is utilized to pursue socio-economic objective, there are three key components of value, namely intrinsic value, institutional value and instrumental value. These three values reflect respectively the internal intellectual and spiritual capacity, organizational techniques that could generate values as well as the capacity of cultural effects used to achieve socio-economic objectives (Donelli et al., 2021: 172–173). Given that cultural activities possess the potential to generate significant profits for the city, their contributions are undeniable as societies are all in need of employment provision and a workable and operational budget with which to keep social harmony.

2.2 *Place Marketing and Re-imaging to Draw in Tourists*

(a) Place Marketing

As suggested earlier, Singapore's urban tourism, due to lack of historic heritage, has to rely on man-made ingredients. As such, place marketing and re-imaging are perceived as essential parts of place destinations to draw in tourists. In this tourism context, the essence of place marketing and re-imaging which can turn man-made sites into marketable place products for touristic promotion to potential customers is addressed (see Page & Hall, 2003). Subsequently, we move on to the Singapore sources of urban tourism where the artificiality issues have arisen.

As a city state which has merged central government and local state administration into one entity, Singapore's strong central state has virtually the full capacity to undertake market-led place marketing strategies or regeneration of its central city area. Regenerative measures in turning the central city slums of the 1960s to a modern city from the 1980s have been a success story (Wong & Yap, 2004). However, this dramatic renewal process had phased out most "old and unsightly" traditional townscape. As Singapore re-establishes itself in its pathway towards modernity suited to tourism, place building and re-imaging have required enormous efforts.

Place marketing requires a relatively heavy initial investment by public authorities or private investors or both to be engaged in creating fresh new sites or improving existing physical resources to enhance their image as a place for visitors. This may be called a place-image commodification process (Law, 1992). In a free market economy, this place-marketing process should be interpreted as an amalgam of two sources of investment. One is the non-profit social marketing where public authorities lay basic infrastructure to support private profit-led ventures whose potential profit-making outcome justifies a taxable revenue return to the government. As private investors, their key concern is the relationship between customer demand and their products. Their products have to be responsive to the needs of the customers, market characteristics and market behaviour. In marketing a city which is spatial in a geographical setting, the efficient mode of product and service delivery has to be made highly accessible where "potential buyers and users are aware of the place's distinctive advantages" (Page & Hall, 2003: 299, Kotler et al., 1993).

Accordingly, Page and Hall (2003: 300–301) have identified five stages of marketing strategy that a local authority should follow, summarized as follows:

(a) Market auditing: A systematic analysis of the city's market position in terms of its external environment containing issues that cannot be influenced by planning and internal organizational interventions. Using commonly the SWOT (Strengths, Weaknesses, Opportunities and Threats) analysis, such external auditing approach will assist in shaping appropriate policies and meeting the needs of the visitors;
(b) Target market identification: This approach is to identify and analyse the spatial behaviour of visitors and their decision-making process on choices of consumption products in the city;
(c) City quality identification: In practice, a complex quantitative multi-criteria evaluation technique would be used to compare the city's products with those of other comparable cities;
(d) Product innovative development: Adjusting and developing new products to meet the needs of the market; and
(e) City image promotion: Promotion efforts are dedicated largely towards the target market, and image building plays a key role.

Cultural sites in the inner cities possess heritage resources that could be used to help create a sense of place as an attractive tourism product. Such product should be well integrated into the overall planning and management objective of the city. Indeed, marketing of urban tourism objects involves two distinct markets: the local residents and visitors.

In real life, local residents and visitors do have conflicts of interest in urban tourism activities (Fainstein, 2011). As research in highly developed economies has revealed, whilst city tourism has the capacity to revitalize post-industrial city centres, the resulting traffic congestion, rising cost of living, real estate speculation, loss of local identity, household dependency of tourism and pollution have in many cases downgraded local lifestyles. Concerns have arisen as to how best to manage the impact of tourism on the recipient end and residents affected (Tokarchuk et al., 2016). Museums, galleries, symphonies, sports activities and exhibitions are local as well as visitor destinations of tourism. Revenues generated, taxes payable and employment opportunities thus created have strong implications to the growth of cities under investigation (Suh & Gartner, 2004; Murphy, 1992).

(b) Place Re-imaging

Public efforts via an institutional system or formal organization do help to promote destination images in tourism. Such efforts have good potential to contribute towards shaping new markets by encouraging investors to invest in tourism business and travelers to cities. City reimaging measures could include building up prestige centres, hosting major international events and fairs to draw in visitors and event organizers (Gotham, 2007; Hall & Hamon, 1996).

Re-imaging of problematic places has become a normative response in many Western cities at the time of deindustrialization or lands at central areas are being revaluated as globalization has intensified from the 1980s forcing manufacturing industries to move overseas or be phased out. A case in point is London's Docklands, an inland port on River Thames.

London Docklands represents a landmark transformation in the late twentieth century where during the 1970s–1980s, over half of the area was characterized by run-down poor housing, and low employment. Half a century later following a series of regeneration campaigns, London Docklands has turned itself into one of London's largest centres of classy riverside properties, a transport hub and a new central business district rich with *in situ* employment. By August 2020, with the launching for a new 5000-home neighbourhood in the Silvertown area of Docklands, the last derelict sites are expected to be removed and the virtual completion of the re-imaging plan a few years later. This characterization of a successful turnover has taken place in a former devaluing inner port area due to its disuse in a changing era. The regeneration strategy in the Docklands did work (Thomas Backlog, 2021; Law, 1992).

The key objectives in regenerating the Docklands were to attract tourism expenditures, in particular the higher budget business spenders, and create employment not only in the tourism industry but in the higher end professionals and white-collar workers. The re-imaging of Docklands has brought along other features contributing to tourism in the form of convention centres, shopping malls, sports complexes, indoor arenas, and development or renewal of leisure and cultural services including museums, art galleries, and hosting of art festivals (see Page & Hall, 2003: 308–309).

3 Singapore's Sources of Urban Tourism

History of Singapore is relatively short, despite historical records of human settlements on its coastal areas as fishermen and sojourners who sailed between South China Sea and the Indian Ocean through the Straits of Melaka. Authentic modern history of Singapore began with European colonization in the Far East in 1819 when Stamford Raffles from the British East Indian Company established the small island as a trading post for business. The island of Singapore was found to be a deep and excellent natural harbour for being close to the Straits of Melaka with ample supply of fresh water, and timber for repairing ships. On arrival in January 1819, Raffles found a small Malay settlement with about 120 Malays and 30 Chinese at the mouth of the Singapore River, headed by *Temenggong*, a senior executive from the Johor Sultanate (Turnbull, 2009).

Raffles seized the opportunity to support the other brother Tengku Hussein of the ruling Johor Sultan to the throne. Raffles offered to recognize Hussein as the rightful Sultan of Johor, and Hussein in return granted the British East India Company the right to establish a trading post on Singapore, and a yearly payment of 5000 Spanish dollars. Consequently, a treaty signed between them in February 1819 justified legally the cessation of rights to the East Indian Company and marked the founding of modern Singapore. The colony quickly grew in population as and when neighbouring traders and population began to flock in. By 1825, the population reached about 10,000 and trade volume 22 million Spanish dollars. Singapore had remained as predominantly a trading post with a major seaport in the Southeast Asian region throughout the whole colonial era of almost 150 years.

Throughout the nineteenth century and the first half of the twentieth century, the trade and capitalist world system built by the British Empire had shaped Singapore's urban cultural landscape infused with both metropolitan and immigrant characteristics. More notably, such features were blended into the colonizers' industrial core and the colonized periphery relationship. In that core-periphery system, capital accumulation in the core relied on the production of commodities and low wage labour in the periphery. The early days of Singapore as the peripheral state with a deep-sea port formed an important part of the British world system production processes in support of its industrialization. On the top of that, there was a small rural hinterland and low-order services in the central area, often of informal economy and "self-employed" in character. Slums and squatters were widespread, and poverty was common place (Turnbull, 2009; Wong & Yap, 2004).

3.1 Heritage Tourism

As a British colony during 1819–1959, Singapore was ruled primarily as an entrepôt, serving the British empire its trade interests in the Far East. During the period of colonial rule, large numbers of immigrants were brought in from China, India and the neighbouring Malay Peninsula and the Archipelago. Multiple cultural heritage of different ethnic backgrounds had formed the backbone of Singapore urban heritage.

As drastic urban renewal took place following independence in 1965, the physically dilapidated urban structure was largely phased out, leaving little city heritage behind. When export-led industrialization took off in the late 1960s, ancillary urban functions such as banking, mass-produced public housing scheme, known as Housing and Development Board, wholesale and retail trade, and expressway network and communications followed suit to support industrialization and financial upgrading. Massive development and redevelopment of the Singapore island has left few stones unturned and numerous frail and "unworthy" heritage structures were inevitably sent to history in the modernization process (Wong & Yap, 2004). Attention is now turned to Singapore's heritage which has been used to promote tourism, despite its limited resources in fulfilling the need.

At the beginning of the 1960s when tourism was promoted as one of the economic pillars to help booster growth, Singapore took the recommendation of the United Nations advisor, Albert Winsemius, that the British imperialist legacy such as Stamford Raffles statute at the mouth of Singapore River ought to be left to stand untouched. Winsemius felt strongly that this would help convince investors from the capitalist West, including Japan that Singapore was politically stable and they were welcome (Centre for LiveableCities, 2015).

Heritage conservation is one way Singapore has applied to preserve the distinctive outlook and character of its rare traditional places with historical and ethnic values. In 1986, the first Conservation Master Plan was released and in 1989 the Singapore Parliament adopted the Planning (Amendment) Act to establish conservation as a planning function. The Act saw subsequently a number of sites gazetted and adaptive reuse of conservation buildings was undertaken as a resilient approach to make traditional buildings fit better to modern use for management and visitor consumption. In other words, the heritage buildings have been converted adaptively to fit in as living areas, and to an extent a living "museum" with commercial values.

There is an opportunity cost by conserving rows of 2- to 3-storey low-rise shophouses for the three ethnic conservation zones in prime city areas (Chinatown, Little India and Kampong Glam) in terms of building height and buildable floor areas. They are also known as "historic districts" or "ethnic quarters". Evolving from lively and dynamic old quarters, albeit chaotic, of the early immigrants, Chinatown at Kreta Ayer downtown area took a sharp physical transformation after its revitalization development since the start of conservation in the early 1990s. In the late 1980s, these diverse ethnic backgrounds represented by three respective historic sites known as Chinatown for ethnic Chinese, Little India for ethnic Indians and Kampong Glam for ethnic Malays were physically established (VisitSingapore.org, 2023). They have been zoned as ethnic tourism sites (Fig. 3.1).

Fig. 3.1 Gastronomy tourism (**a**) Indian cuisine, Little India (**b**) Chinese restaurant, Chinatown (**c**) Malay cuisine, Kampong Glam Conservation Area. (Source: Photos by author taken in October 2023)

3.2 Artificial Tourism

Left with little but the choice of artificial tourism, Singapore tourism Board has dedicated itself to create a specific tourism and make it a rationalized industry with investment capital devoted to "aestheticization of local culture and the production of spaces of consumption, leisure and entertainment" (see Gotham, 2007: 306), despite its artificial character. Arguably, tourism development does not restrict itself to building resorts and hotels and their related services alone, it requires essentially the support of infrastructure in both physical infrastructure and infostructure. Airport, highways and communications lines are just other supporting elements. From the 1980s, the city-state has launched a series of information and technology plans to move itself into an international financial and technological hub (Wong, 1999).

The most critical move was the decision to build a world class airport at the eastern end of the island in the early 1970s at Changi from reclaimed land, as by then already over 90% of Singapore tourists arrived by air. Changi Airport's scale and carrying capacity has later proved to be highly functional, efficient and made Singapore proud as initially set as its objectives (Centre for LiveableCities, 2015: 24). Tourist arrivals at Changi Airport reached its pre-Covid peak in 2019 at 68.3 million, a 13-fold volume of Singapore's population, characterized by a total of 382,000 commercial flights taking off or landing in the airport (Kok, 2023).

As one of the key business hubs in Asia, Singapore by the early 2000s had gained good ground in making itself a destination for resorts. It was then publicly acknowledged that business events in destinations must be accompanied with exciting leisure options such as resorts and recreational sites (Chua, 2005). Sentosa Island (previously known as Belakang Mati and used as a British military base), situated immediately south of the Singapore main island, turned out to be an ideal site to be developed into a recreation and tourism resort (Fig. 3.2).

As globalization and business tourism intensified in the 1990s, business destinations for Meetings, Incentives, Conferences and Exhibitions (MICE) were subsequently emphasized. Suntec City in the Marina Downtown Core in Singapore is symbolic of MICE development and resort development. The City was constructed in phases and the last building was completed in 1997 with a mixed-use of shopping mall, offices and a convention centre, spreading out in five buildings. Suntec City captured international media attention by holding the inaugural World Trade Organization (WTO) ministerial conference during December 9–13, 1996. Another conventional centre which came into existence in 2010, known as Marina Bay Sands, added to the list as Asia's leading destination for business, leisure and entertainment. Here, there are 250 meeting rooms and 24 ballrooms to meet challenges in simultaneously organizing a series of international conferences and events (Marina Bay Sands, 2023).

Not surprisingly, there is a casino set up in the gigantic building, a decision made to allow public gambling in contrast to Singapore's earlier moral verdict not

Fig. 3.2 Artificial tourism (**a**) Gardens by the Bay, Downtown Singapore. (**b**) Six-star Golden Sands Hotel, Downtown Singapore. (**c**) Sentosa Resorts World, Sentosa Island. (Source: Photos by author in November 2023)

to have it. Since its independence in 1965, Singapore had banned casinos on a very firm stance (Henderson, 2012: 141). To justify Casino's existence, the Prime Minister of Singapore, Lee Hsien Loong in 2005 argued the importance of taking part in the globally competitive business environment where global cities were reinventing themselves to make themselves relevant. Should Singapore choose not to be involved, he said, Singapore would be losing ground in tourism, and after all, casino was only a small component of the integrated resort but it helped Singapore to reinvent itself (Singapore Parliament Reports, 2005).

Since the introduction of the integrated resorts and the casinos in 2010, there had been business success and report of desired economic returns. The two casinos at Marina Bay Sands and Sentosa had come strong with estimated gross gaming revenue of US$6 billion, joining hence Macau as a second Asian gaming centre, contributing to the city-state's desired paths to growth (Henderson, 2012; Cohen, 2019).

3.3 Weaknesses of Artificial Tourism

Singapore revenue-oriented tourism may be termed a commodified consumer cultural tourism. However, the city-state's devotion to artificial aestheticization of touristic objects has on one hand commodified marketable products, yet has executed by doing so with a strong rationalization supplemented by efficiency in its development of urban tourism. Rationalization in this respect covers heavy investment, publicity and place promotion efforts, but an excessive tourism promotion through marketing could undermine authenticity resulting in ingenuine outlook (Fainstein, 2011).

Man-made objects could be highly successful in drawing crowds but they lack somehow thickness as intangible cultural heritage which require time and accumulated human input to build up its depth. By UNESCO's definition dated 2003, intangible cultural heritage indicates:

> [T]he practices, representations, expressions, knowledge, skills - as well as the instruments, objects, artefacts and cultural spaces associated therewith - the communities, groups and, in some cases, individuals recognize as part of their cultural heritage. This intangible cultural heritage, transmitted from generation to generation, is constantly recreated by communities and groups in response to their environment, their interaction with nature and their history, and provides them with a sense of identity and continuity, thus promoting respect for cultural diversity and human creativity (Cited in You & Hardwick, 2020: 4).

Singapore's casinos, Formula one Grands Prix, the resorts and theme parks are business undertakings which do not transmit, produce and reproduce local traditions but are merely a form of expressive commercial culture designed for more practically international tourists. Local communities are largely not targeted as the key clientele, as foreigners are more seen as the source of revenue generation.

Man-made touristic parks have also resulted in failure in Singapore. Some were phased out due to rise in land values and intensification was needed while some others simply could not attract visitors. To cite a few, the former category includes New World Amusement Park (1923–1987), Great World Amusement Park (1929–1978), Gay World Amusement Park (1936–2000), Wonderland Amusement Park (1969–1988) whereas the latter covers Tang Dynasty City (1992–1999), Escape Theme Waterpark (2000–2011), Fantasy Island, Sentosa (1994–2002) (see TheSmartLocal, 2023).

3.4 The Casino Debate

Gambling is secularly and religiously a controversial issue. Adverse consequences are multiple such as crimes associated with money laundering, drug trafficking, prostitution, alcoholism, personal bankruptcy and other hidden cost resulting from medical treatments. Casino was once cited by the Australian Prime Minister Paul Keating as "ugly as sin" (Henderson, 2012; Hall & Hamon, 1996: 34). Despite awfully sad stories which we have often heard of, casinos are popular in many countries and they act as an attractive means of leisure to tourists who find thrill of a possible reward. The probability of winning in casino for those taking part is however lower than the casino operators who make their living from it. Indeed, the house always wins based on the return on casino investments (Cohen, 2019). As a whole, addicted gambling is highly detrimental to social well-being of a community.

In Australia, casino development has been largely regarded by state governments as a way to improve destination image as well as a strategy to promote city marketing and as a means of city governance. More particularly for Australia, as a welfare state perhaps it is critical to increase the revenue base to help pay off city debts and subsidized city projects. Needless to say, casinos do have the potential to serve as a rejuvenation measure to help rebuild rundown inner city sites (Hall & Hamon, 1996).

As a well-developed economy after four decades of modernization which moved from labour intensive industrialization to capital-intensive export-led cum knowledge-based economy, Singapore's political economy and globally linked structure has consolidated a step further towards deindustrialization like most of the developed Western economies. High-end services inclusive of gaming has been seen as a tool to revitalize and bolster growth of the city-state. Casinos have thus been developed within a broader integrated scope of redevelopment and reimaging. At this stage, Singapore would need to include casinos as a commodity product to maintain its growth dynamics which are increasingly difficult to achieve as a developed city-based economy (see Hall & Hamon, 1996).

4 Conclusion

Artificial touristic features complement genuine heritage attractions when the latter are in short supply. This combination has proved to be a success as it is the case in Singapore. However, artificiality which is derived from human acumens may be misjudged and incompatible in place remaking or reimaging. In terms of design, artificial features may fit in well physically but they could have created missing links in historical and cultural continuance.

The reshaping and revitalization of Singapore's Chinatown into a "theme park" has been controversial as there is concern that the consequence could be an "inauthentic and contrived Orientalist theme park for tourists, instead of organic development that catered to the needs of the local community" (Tan, 2001: 9). Public remarks like this had exerted pressure to Singapore Tourism Board to withdraw its earlier "theming" idea on Chinatown. The Board is now more cautious not to "over theme" traditional heritage merely for the sake of tourist attractions but to share public sentiments and to attach more to local cultural roots.

In the first place, the Chinatown of the old that existed prior to independence filled with noises of yelling hawkers and bargaining customers and filthy waste scattered all over the wet market cannot be recreated in a strictly enforced cleanliness standard set by the present Singapore government. Certain preserved objects or physical structures have been well protected and renovated in accordance with conservation norms but the spirit and ambience of the old streetscape could never be revitalized. Despite place-making efforts made by the officially appointed Chinatown Business Association in 1999 to spearhead rejuvenation of the historical area, there has been limited success.

Despite negative impacts, urban tourism as a whole has shown enhanced quality of life, characterized by the rise in local incomes, improved public transport, and more resources dedicated to education and school infrastructure, as it is the case in China in general (Tokarchuk et al., 2016). In Singapore, over the past half of a century, the city-state's tourism commodities have been systematically set up, and sold in the market. Those tourism-related premises or sites and their related services have been spatially fixed and consumed as saleable products by both domestic residents and international travellers. More particularly, as a global business hub, this fixity of place which is a non-transferable product requires non-local tourists to make the commodification process effective. As it is a substantial revenue generator in cultural tourism activities, it is seen as a key economic contributor with a high significance of "localized cultural valorization" (see Gotham, 2007).

The nature of Singapore's state capitalism characterized by acts of an effective entrepreneur has a strong orientation towards building profit-led government-linked companies (GLCs). These GLCs are run on principles of market forces and are managed by highly paid selected elites who are efficient in performing their duly tasks (Volgyi, 2019). Given this nature of profit-orientation, Singapore tourist agencies are resilient and could be restructured to fit in new circumstances to counter challenges. As such, the contents of tourism, like other economic sectors in Singapore which have been consistently updated and restructured, would undergo new phases of change in the near future. Yet who knows?

References

Cardenas, D. A., Byrd, E. T., & Duffy, L. N. (2015). An exploratory study of community awareness of impacts and agreement to sustainable tourism development principles. *Tourism and Hospitality Research, 15*(4), 254–266.

Centre for LiveableCities. (2015). *Planning for tourism: Creating a vibrant Singapore*. Singapore.

Chua, J. (2005). Resorts could double visitor arrivals, add $1.5b to GDP; property players hope construction industry will be revived. *The Business Times*, April 19, p. 4.

Cohen, M. (2019). *Singapore makes casinos an offer they can't refuse*. https://www.forbes.com/sites/muhammadcohen/2019/04/09/singapore-makes-casinos-an-offer-they-cant-refuse/?sh=2fe8158a5279

Donelli, C. C., Trimachi, M., Pratici, L., & Fanelli, S. (2021). The value of culture in building resilience in cities. In K. Lehman, I. Fillis, & M. Wickham (Eds.), *Exploring cultural values: Contemporary issues for theory and practice* (pp. 171–182). Emerald Publishing.

Edwards, D., Griffin, T., & Hayllar, B. (2008). Urban tourism research: Developing an agenda. *Annals of Tourism Research, 35*(4), 1032–1052.

Fainstein, S. S. (2011). Review of book: Urban tourism and urban change: Cities in a global economy by Costas Spirou. Routledge, New York. *Contemporary Sociology, 41*(1), 104–105.

Fox, R. (2023). *Urban culture*. https://www.britannica.com/topic/urban-culture
Gotham, K. H. (2007). Destination New Orleans: Commodification, rationalization, and the rise of urban tourism. *Journal of Consumer Culture, 7*(3), 305–334.
Hall, C. M., & Hamon, C. (1996). Casinos and urban redevelopment in Australia. *Journal of Travel Research, 34*, 30–36.
Henderson, J. C. (2012). Developing and regulating casinos: The case of Singapore. *Tourism and Hospitality Research, 12*(3), 139–146.
Kok, Y. F. (2023). Passenger traffic at Changi Airport expected to recover fully by 2024 or possibly earlier. *The Straits Times*, March 3, 2023.
Kotler, P., Haider, D. H., & Rein, I. (1993). *Marketing places: Attracting investment, industry, and tourism to cities, states and nations*. The Free Press.
Law, C. M. (1992). Urban tourism and its contribution to economic regeneration. *Urban Studies, 29*(3/4), 599–618.
Lehto, X. Y., & Lehto, M. R. (2019). Vacation as a public health resource. Toward a wellness centered tourism design approach. *Tourism and Hospitality Research, 43*(7), 935–960.
Marina Bay Sands. (2023). *Meetings and conventions*. https://www.marinabaysands.com/expo-events-and-convention-centre/meetings-and-conventions.html
Murillo, J., Vaya, E., Romani, J., & Surinach, J. (2013). How important to a city are tourists and day-trippers? The economic impact of tourism on the city of Barcelona. *Tourism Economics, 19*(4), 897–917.
Murphy, P. E. (1992). Urban tourism and visitor behavior. *American Behavioral Scientist, 36*(2), 200–211.
Page, S. & Hall, C. M. (2003). *Managing urban tourism*. Prentice Hall, Harlow.
Pearce, D. (2001). An integrative framework for urban tourism research. *Annals of Tourism Research, 28*(4), 926–946.
Singapore Department of Statistics. (2022). *Tourism direct contribution as share of Singapore GDP*. http://statista.com/statistics/1079842/singapore-tourism-direct-contribution-to-gdp/
Singapore Parliament Reports. (2005). Proposal to develop integrated resort, statement by the Prime Minister and Minister for Finance, 18 April Vol. 80, Col. 61. http://sprs.parl.gov.sg/search/report.jsp?currentPubID=00004697-WA
Singapore Tourism Board. (2020). Overview. https://www.stb.gov.sg/content/stb/en/about-stb/overview.html, and Achieving Quality Tourism. https://www.stb.gov.sg/content/stb/en/about-stb/achieving-quality-tourism.html
Suh, Y. K., & Gartner, W. C. (2004). Perceptions in international urban tourism: An analysis of travelors to Seoul, Korea. *Journal of Travel Research, 43*, 30–36.
Tan, T. K. (2001). *Social capital and state-civil society relations in Singapore*. Institute of Public Studies.
TheSmartLocal. (2023). *9 Forgotten Theme Parks of Singapore's past that we wish were still around today*. https://thesmartlocal.com/read/old-theme-parks/
Thomas Backlog. (2021). *How will 2021 transform the Docklands?* https://www.hastingsinternational.com/content/news/244-docklands-regeneration-2021-how-will-2021-transform-the-docklands
Tokarchuk, O., Gabriel, R., & Maurer, O. (2016). Development of city tourism and well-being of urban residents: A case of German magic cities. *Tourism Economics, 23*(2), 343–359.
Turnbull, C. M. (2009). *A history of modern Singapore 1819–2005*. NUS Press.
United Nations. (2022). *Singapore: International visitor arrivals 2021*. http://www.statista.com/statistics/977993/total-international-visitor-arrivals-singapore/
UNWTO. (2019a). *Urban tourism*, https://www.unwto.org/urban-tourism
UNWTO. (2019b). *8th UNWTO global summit on urban tourism "smart cities, smart destinations"*. 9 Oct 2019, https://www.unwto.org/8-unwto-global-summit-on-urban-tourism-smart-cities-smart-destinations
VisitSingapore.org. (2023). *Ethnic quarters*. https://www.visitsingapore.org/explore-singapore/ethnic-quarters/. Retrieved 30 May 2023.

Volgyi, K. (2019). A successful model of state capitalism: Singapore. In M. Szanyi (Ed.), *State ownership in the varieties of capitalism* (pp. 275–296). Central European University Press.
Wong, T.-C. (1999). The transition from physical infrastructure to infostructure: Infrastructure as a modernising agent in Singapore. *GeoJournal, 49*, 279–288.
Wong, T. C., & Yap, L. H. A. (2004). *Four decades of transformation: Land use in Singapore 1960–2000*. Marshall Cavendish International.
You, Z. Y., & Hardwick, P. A. (2020). Guest editor's introduction. *Asian Ethnology, 79*(1), 3–19.

Tai-Chee Wong has his BA and MA degrees in Urban and Regional Planning from the University of Paris and earned his PhD in Human Geography from the Australian National University. After teaching 20 years at Nanyang Technological University, Singapore, he retired in 2013. He then worked as Distinguished Professor for two years at Guizhou University of Finance and Economics, China, and as Dean and Professor at the Southern University College, Johor Malaysia, from March 2017 to December 2018, and as Deputy Vice-Chancellor at Berjaya University College, Kuala Lumpur from October 2022 to September 2023. He was a Visiting Professor at the University of Paris (Sorbonne IV) and a Visiting Fellow at Peking University and Tokyo University. His main research interests are in urban and economic issues and, more recently, in Malaysian politics. Besides his 16 self-authored and edited book volumes, he has written over 100 academic articles and published widely in international journals including tourism papers.

Chapter 4
Inclusion of Persons with Disabilities in Heritage Tourism Sites in Malaysia

Meg Milligan, Lee Ting Lew, Suzanne Tew-Washburn, and Mary Kristin Medina

1 Introduction

Persons with disabilities (PWD) form the largest minority group in the world but are underrepresented in many activities, including tourism, and the numbers are much higher if persons with temporary disability, such as short-term use of assistive devices following an injury, are included. Globally, the World Health Organization estimates there are 1.3 billion people with significant disabilities, which is 16% of the total population or 1 in 6 people worldwide (WHO, 2023). The exact number of PWDs is challenging to quantify because there are no universally accepted definitions of disability as they vary by location and perspective. Disability is an "umbrella term" for illness and functional limitations that restrict full participation in individual choice activities (Marsin et al., 2014). Definitions include the Americans with Disabilities Act (ADA) (U.S. Department of Labor, 1990) and the biopsychosocial model of the World Health Organization's (WHO) Classification of Functioning, Disability, and Health.

The ADA "prohibits discrimination against people with disabilities in several areas, including employment, transportation, public accommodations, communications and access to state and local government' programs and services" (U.S. Department of Labor, 1990). The United Nations' 2030 Agenda for Sustainable Development (United Nations, 2023) reiterates these goals. Definitions are reflected in laws and public policies, which change as cultures change.

Definitions and connotations are essential to understanding a location's accessibility and inclusion practices. For example, historically, there have been

M. Milligan (✉) · S. Tew-Washburn · M. K. Medina
Troy University, Phenix City, AL, USA
e-mail: mmilligan@troy.edu; stew-washburn@troy.edu; mmedina@troy.edu

L. T. Lew
J&T Express, Kuala Lumpur, Malaysia

T.-C. Wong et al. (eds.), *Cultural Tourism in the Asia Pacific*,
https://doi.org/10.1007/978-3-031-63459-8_4

fundamental shifts in definitions in the United States (U.S.). These are described by Frances and Silvers (2016) and summarized here. Until the late 1800s, the concept of disability emanated from restrictive laws (e.g., married women could not manage property or vote). These laws "disabled" people in the sense that they were not allowed to engage in certain behaviors. This shifted to seeing disability through a medical lens, reflecting a change from loss of rights to loss of function. During the twentieth century, this medicalization was exhibited by a focus on correcting societally perceived defects in persons deemed disabled (such as cochlear implants and vocational rehabilitation programs), compensating others based on a medically determined inability to work, and segregating some people in institutions (such as institutions for the blind and insane asylums). The word "handicap," pejorative by today's Western standards, was used, and vestiges persist in the English language in terminology such as "handicapped parking." The handicap designation is mired in the pseudoscience of eugenics promulgated by Charles Darwin's cousin, Sir Francis Galton (1822–1911). It denotes defective and disadvantaged and reflects external social judgments of an individual's value (Burgdorf & Burgdorf Jr., 1975).

Recently, there has been a shift toward individual agency in deciding the terminology used. Person first (e.g., persons with disabilities) or disability first (e.g., disabled persons) language is a matter of personal preference; however, person first seems more widely used (Iezzoni, 2016). Other approaches to language that reduce the binary categorization of people (e.g., disabled-not disabled) are proposed, such as neurodiverse and differently-abled (Austin & Pisano, 2017). Neurodiverse includes everyone, neurotypical and neurodivergent, which normalizes the terminology and avoids the pitfalls of the medical model, such as defining neurotypical and ignoring positive attributes of disability (Loo et al., 2021). The tourism industry tends to overlook neurodivergence, and addressing neurodivergent needs could expand this sector by attracting many more tourists (Jepson et al., 2023).

Twenty-first-century thinking emphasizes the individual's intrinsic value and equal rights to an unrestricted environment regardless of ability. This aligns with inclusion. The 2020 Inclusiveness Index: Measuring Global Inclusion and Marginality, published online by the Othering & Belonging Institute of the University of California at Berkeley, defines inclusivity "by focusing primarily on the performance of groups that span salient social cleavages, such as gender, race, ethnicity, religion, sexual orientation, and (dis)ability," and this is foundational to Universal Design (UD). As the Center for Universal Design at North Carolina State University (1997) explains, "If design can marginalize people and push them to the wayside, it can also unite people and bring them together. Universal design is an approach to design that strives to make our products and environments usable for people to the greatest extent possible without the need for adaptation or specialized design." This philosophy significantly differs from the typical design that targets the average ("neurotypical") person and expects other individuals to adjust.

Social inclusion (Martin & Cobigo, 2011) and community inclusion (Malijani, 2019) are other vital aspects of the right to full participation in the tourism sector. "Social inclusion is about making sure all individuals are able to participate as valued, respected and contributing members of the society" (Genç & Genç, 2017: 194).

Access to resources is a determinant of social inclusion or exclusion. Community inclusion focuses on equitable allocation of the benefits of tourism. Accessibility contributes to social and community inclusion.

Additionally, there are well-established links between leisure and life satisfaction (Sirgy et al., 2017). For example, a study conducted in South Korea found that employed persons with disabilities who were satisfied with their leisure activities reported higher overall life satisfaction than those who reported dissatisfaction with their leisure (Kim et al., 2021). Another study showed that adolescents with various disabilities had greater satisfaction when they self-managed their leisure pursuits (Doistua et al., 2020). This research, among other studies, highlights the importance of access to enjoyable leisure activities to improve the quality of life for people with disabilities, and this is another reason to expand accessibility to cultural heritage sites.

If tourism is to include people with disabilities, a best practice is to use UD principles that accommodate the larger population of individuals with and without disabilities. The UD approach not only serves more individuals but, because it is adopted at the beginning of construction, it is much more economical than trying to construct accommodations for people with disabilities at a later time. Retrofitting is costly and challenging but unavoidable for many heritage sites.

The absence of uniformity in accessibility procedures makes standardization difficult. The International Organization for Standardization's ISO Standard 21902:2021 (2021) incorporates UD policies and provides some context for developing specific objectives to address accessibility in tourism internationally. The ISO Standard 21902:2021 does not provide a specific structured plan in the traditional sense; however, it offers a valuable framework for organizations and individuals in the tourism industry, including recommendations for managing protected areas (UNWTO, 2023).

Improving the accessibility of tourism services requires taking a broad-based approach that extends accessibility to include transportation and accommodations. Visionary planning based on the ISO 21902:2021 framework should include areas such as assessment and awareness, training, product and service improvement, customer service optimization, incentives, budgeting, accessibility considerations, and continuous monitoring. A carefully designed approach to inclusion aligned with ISO 21902:2021 ultimately promotes high-quality, accessible experiences. Individuals and companies with proper training are encouraged to implement accessibility measures while saving on construction costs by incorporating accessibility into the planning stage (UNWTO, 2023).

Inclusion through accessibility has moral, economic, and legal justifications. These are addressed from larger-scale perspectives, such as the United Nations Convention on the Rights of Persons with Disabilities (United Nations, 2006) and the International Classification of Functioning, Disability, and Health by the WHO (2021) to more localized perspectives, such as national. Worldwide entities are increasingly influential as globalization, the process by which "cultures influence each other and become more alike through trade, immigration, and the exchange of ideas" (Arnett, 2002: 774) continues, cognizant of the tension between globalization and the nation-state (Smallman & Brown, 2020).

This chapter examines inclusion and accessibility as basic global aspirations within the overarching goal of universal design. We examine one of the United Nations Educational, Scientific, and Cultural Organization (UNESCO, n.d.) World Heritage Sites in Malaysia, Kinabalu Park, as an applied example in Asia. This site is more likely to exemplify inclusion due to its well-recognized significance and identification with UNESCO, cognizant of the challenges inherent in increasing inclusion in nature-based tourist sites.

UNESCO reviews applications from locations worldwide as potential additions to the list of World Heritage Sites, including natural land and underwater sites. As of September 2023, there were 1199 UNESCO World Heritage Sites in 168 countries. This prestigious list confers benefits, including attracting tourists, which has socio-economic advantages for those locations. This chapter ends with a conclusion and recommendations for improvement.

2 Accessibility, Inclusion, and Universal Design

"Broadly defined, accessibility refers to the ability of an individual to enter, navigate, comprehend, and use features of an environment, including physical, digital, and conceptual environments" (Harniss, 2014: 6), In its International Classification of Functioning, Disability, and Health, the WHO (2017) reiterates the importance of participation in social activities as a fundamental human right. This denotes a barrier-free environment for all people regardless of ability or setting. This is aspirational, and more pragmatic definitions are provided in regulations, guidelines, statutes, and laws, such as ADA and ISO Standard 21902:2021.

Inclusion is another general term that has a specific meaning when applied to tourism, namely, inclusive tourism, which Scheyvens and Biddulph (2017: 4) define as "transformative tourism in which marginalized groups are engaged in ethical production or consumption of tourism and the sharing of its benefits." PWDs are considered a marginalized group and a key stakeholder. Accessibility facilitates equitable "consumption of tourism."

The Convention on the Rights of Persons with Disabilities (CRPD, 2006) and its Optional Protocol AR/RES/61/106, is a tool deployed by the United Nations to help promote, protect, and ensure that all persons with disabilities have equal human rights and fundamental freedoms without discrimination. It went into effect in 2008. A UN committee created this document within an unprecedented eight sessions between 2002 and 2006, "making it the fastest negotiated human rights treaty" (United Nations, n.d.). The CRPD, a disability rights treaty of the United Nations, demonstrates a broad base of commitment to people with disabilities.

The CRPD defines a person with disabilities as having long-term physical, mental, intellectual, or sensory impairments that create barriers keeping them from fully or effectively participating in a society equal to those without a disability. The CRPD includes 50 articles that use a UD model, with the first 30 articles about different categories of rights for persons with disabilities. In contrast, women and children with disabilities each have their own article section. Additionally, the article sections explain people's rights and how they will be protected (United Nations, 2006).

Ratification, although important, is just a first step. To be effective, anti-discrimination laws that coincide with the ambitious goals, spirit, and intentions of the CRPD must be established. Each section of the CRPD (such as Article 19 below) covers an important topic. Equality, accessibility, right to life, freedom from inhumane treatments, freedom of expression, education, employment rights, and international cooperation are examples.

The Committee on the Rights of Persons with Disabilities (Article 19) (CRPD, 2014) maintains that people with disabilities have natural rights that command respect for the individual's dignity, autonomy, and independence, as well as the right to full inclusion in all aspects of life afforded individuals without disabilities, including the right to be able to live independently and choose their place of residence and have the right to access in-home, residential, and other community-based support services that aid in their independent living and help to prevent isolation. Community services and facilities that are used by the general population should be available on an equal basis to people with disabilities. This includes access to cultural heritage sites.

Twenty-first-century thinking emphasizes the individual's intrinsic value and equal rights to an unrestricted environment regardless of ability. This aligns with inclusion. The 2020 Inclusiveness Index: Measuring Global Inclusion and Marginality, published online by the Othering & Belonging Institute of the University of California at Berkeley, defines inclusivity "by focusing primarily on the performance of groups that span salient social cleavages, such as gender, race, ethnicity, religion, sexual orientation, and (dis)ability," and this is foundational to Universal Design (UD). As the Center for Universal Design at North Carolina State University explains, "If design can marginalize people and push them to the wayside, it can also unite people and bring them together. Universal design is an approach to design that strives to make our products and environments usable for people to the greatest extent possible without the need for adaptation or specialized design" (The Center for Universal Design | Universal Design Resource). This philosophy significantly differs from the typical design that targets the average ("neurotypical") person and expects other individuals to adjust.

3 Inclusion in Asia

In its policy statement on its Asian and the Pacific Decade of Disabled Persons from 1993 to 2002, the United Nations Economic and Social Commission for Asia and the Pacific (UN-ESCAP, 1993) focused on several areas, including legislation, policies, and programs that convey issues that people with disabilities may encounter. Each section has a list of concerns that can be developed into policies to help support the equality for people with disabilities in Asia and the Pacific, such as improving living conditions in rural areas, access to rehabilitation services, assistive devices, education, training and employment, and freedom of movement on roads and public transport systems in urban areas.

UN-ESCAP's statement promoted the rights of persons with disabilities in Asia and the Pacific through its efforts to implement the Incheon Strategy. The Incheon Strategy to "Make the Right Real" takes a holistic approach to disability-inclusive development, providing disability-specific development goals, targets, and indicators. Building on the CRPD and more than 2 years of consultations between governments and organizations, the Incheon Strategy has enabled the Asia and Pacific region to track progress toward improving the quality of life of persons with disabilities.

The Sustainable Development Goals (SDG) underlying the Incheon Strategy provide the world with a universal, people-centred, and inclusive approach to global development until 2030. The Incheon Strategy and the SDG 2030 are based on respect for human rights and take a people-centred and gender-sensitive approach to development. The SDG Summit in 2019 set the agenda for Sustainable Development Goals 2030. SDG 16: Promote peaceful and inclusive societies for sustainable development and provide access to justice for all (Altamas, Institute of Human Development 2019 Review of Evidence on Disability Programmes in South Asia). In Asia and the Pacific, governments, civil society, and self-advocates with disabilities can leverage the synergies between the CRPD, the Incheon Strategy, and the SDGs to ensure that persons with disabilities are at the forefront of this historic movement to transform our world.

Asia has tremendous cultural diversity across its 48 nations, and it is difficult to justify many statements about Asian culture in general. One general distinction is individualism versus collectivism (Heine, 2020). Each has a characteristic self-concept that shapes attitudes and behavior. Individualistic cultures focus on individual responsibility and self-sufficiency, fostering an independent self-concept. Western cultures tend to be individualistic. Collectivist cultures emphasize group well-being and ingroup affiliation, encouraging an interdependent self-concept. Asian cultures tend to be collectivist. Each view has an inherent rationale for inclusiveness as an individual right or concern for the community. This collectivist orientation can be used to reduce stigma and garner support for projects targeting greater inclusiveness.

There are vestiges of stigmatizing cultural views in Asia, such as physical disability may be seen as punishment for misdeeds in a prior life, possession by evil spirits, or it is a family disgrace (Parker, 2001; Saravanabhavan & Saravanabhavan, 2001). Stigma is universal and developed through natural selection, the driving force of biopsycho (biological and psychological) coevolution. Disability denotes a negative deviation from normative ability, which can trigger avoidance as a survival-enhancing and health-protective mechanism. Associated attitudes, which are aspects of the environment, are barriers to the full participation of persons with disabilities. This is part of the human behavioral immune system, the psychological equivalent of the biological immune system (Murray & Schaller, 2016). Humans innately prefer associating with people they perceive as part of their in-group who share commonalities and are similar to them and avoid those seen as the out-group. Persons with disabilities, especially with physical disabilities, who are readily identifiable as different, are more likely to be targets of avoidance, prejudice, and discrimination. Viewing disability from a medicalized disease model exacerbates this tendency. These human tendencies affect participation in tourism and partly explain the low compliance with accessibility regulations.

4 Inclusion in Malaysia

Malaysia heavily depends on tourism to generate revenue for the nation and local governments. "In Malaysia, the tourism industry is considered the prime contributor to the development of the country's economy" (Balasubramanian & Hanafiah, 2022: 1). Expanding access to heritage sites, cultural and natural, fits this role. Tourism in Malaysia has been declining, possibly due to competition from other Asian nations (Aljohani & Choy, 2021). This might be an impetus for improving accessibility and tapping into this underserved market. Other causes might be a global economic slowdown, political quandaries, travel warnings, and the recent pandemic.

The Malaysia Federal Constitution (1957) established a foundation for citizens' equal rights. This document declared that "'there shall be no discrimination against citizens on grounds of religion, race, descent or place of birth'" (Kamarudin et al., 2013: 2). Although persons with disabilities are not listed, inclusive intent is apparent. Accessibility was mandated in 1991 by the Malaysian Standard Codes of Practice for Disabled Persons (MS) as presented in the Uniform Building (Amendment) By-Laws (UBBL). However, compliance was problematic (Kamarudin et al., 2012). Other notable legal documents include Malaysian Standard 1183: 1990 Code of Practice for Means of Escape for Disabled Persons, the Uniform Building (Amendment) By-Laws (UBBL) 1991, Malaysian Standard 1184: 2002 Code of Practice on Access for Disabled Persons to Public Buildings (First Revision), Malaysian Standard 1331: 2003 Code of Practice for Access for Disabled Persons Outside Buildings (First Revision), and Malaysian Standard 1184: 2014 Universal Design And Accessibility In The Built Environment (Kamarudin et al., 2012).

Other laws addressed heritage such as The National Heritage Act, Act 645, 2005 (Laws of Malaysia, 2006: 18, 35), through which Malaysia identifies, lists, and regulates its cultural heritage. The appointed Commissioner of Heritage is directed "to supervise and oversee the conservation, preservation, restoration, maintenance, promotion, exhibition and accessibility of heritage" (18). In addition, "the owner of a heritage site shall ensure that the heritage site is always in a state of good repair" (35). Combining these dictates with disability accommodation mandates, and "a state of good repair" includes maintaining access for people with disabilities.

Supplementing domestic efforts, Malaysia was one of the 164 signatories of the Convention of Rights of Persons with Disabilities (CRPD) in 2006 (United Nations, 2006). This led to the Persons with Disabilities Act of 2008 (Act 685), which "recognizes disability as an evolving concept and a result of interaction between persons with disabilities (those who have long-term physical mental, intellectual, or sensory impairments) with attitudinal and environmental barriers [that] may hinder their full and effective participation in society" (Marsin et al., 2014: 2). By 2010, no significant progress had been made toward removing barriers, and enforcement was "weak." (Ding, 2010). Furthermore, Abdullah et al. (2017) identified gaps in this legislation that reduced its effectiveness. Addressing these gaps could improve compliance in Malaysia.

Another challenge is resolving tensions between the stipulations in The National Heritage Act of 2005, which states, "No person shall, without the written approval of the Commissioner, transfer, demolish, remove, alter, renovate, export, add to or deal with any National Heritage except in case of urgent and immediate necessity for the safety of persons or property" (Laws of Malaysia, 2006: 69) and the priorities of the Persons with Disabilities Act. Another consideration is that too many alterations could jeopardize a World Heritage site's designation.

A number of research studies have assessed accessibility in Malaysia. For example, Jamaludin and Kadir (2012) looked at the accessibility of three major tourist attractions in Malaysia (Berjaya Times Square, Museum of History and Ethnography Malacca, and Central Market Kuala Lumpur), and noted that the older buildings were less accessible than the newer buildings.

More recently, Zahari et al. (2020) studied wheelchair accessibility to National Heritage Buildings in Malaysia and found significant barriers for people who use wheelchairs, primarily from budgetary constraints. They also noted the age of these buildings, difficulties retrofitting within legislative guidelines, and wheelchair users feeling unwelcomed as additional challenges. Majdi (2021) studied the accessibility of toilets in a sample of public facilities and found that the transportation sector scored lowest on this requirement, followed by tourism, while healthcare and business/retail were better prepared. Notable weaknesses were the height of the toilet paper dispenser and the lack of emergency alarms in accessible toilet booths. In addition, Malaysia scored low on the 2020 Global Inclusion Index (Othering & Belonging Institute, 2020).

Public heritage site structures are required to comply with the Malaysia Standards (MS) and code of practices. The latest version is MS 1184: 2014 Universal Design and Accessibility in the Built Environment—Code of Practice (Second revision). This standard applies to buildings to which the public has access, such as religious buildings, universities, restaurants, swimming pools, and car-parking buildings. It provides the requirements and recommendations related to the principal human abilities that should be considered when designing, constructing, and managing the built environment, from the constructional aspects of access to buildings to circulation within buildings to egress from buildings in the ordinary course of events, and evacuation in the event of an emergency. When fully implemented, this standard is expected to be beneficial to all human beings, including people with mobility impairments, visual impairments, hearing impairments, cognitive or learning impairments, hidden impairments (such as mental, stamina, strength, and allergy), and diversities in age (Rahim, 2021).

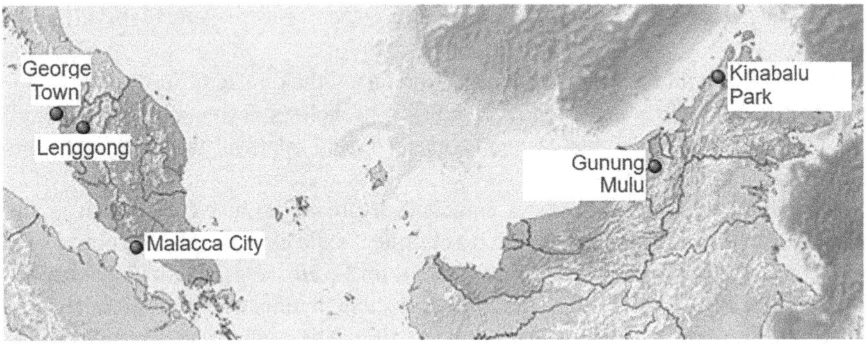

Fig. 4.1 Location of UNESCO World Heritage Sites in Malaysia. (Source: Wikimedia Foundation, 2023)

Although Malaysia has five sites, UNESCO designates them as four World Heritage Sites: 1. Kinabalu Park in Sabah, 2. Malacca and Georgetown, 3. Gunung Mulu National Park, and 4. Archeological Heritage of the Lenggong Valley) (UNESCO, 2024). In 1972, UNESCO initiated this program by enacting a treaty called the World Heritage Convention, whose mission was to preserve cultural locations and promote the conservation of nature (Fig. 4.1).

We describe Kinabalu National Park, its history, and its application to be included as a UNESCO World Heritage Site, followed by observations of its accessibility to people with disabilities. Kinabalu Park is an example of nature-based tourism, also called nature tourism and natural tourism. This type of tourism presents inherent challenges for resource preservation versus economic development and accessibility compliance.

The Seventh Malaysia Plan (1995–2000), the Eighth Malaysia Plan (2001–2005), and the Ninth Malaysia Plan (2006–2010) were pivotal in encouraging nature-based tourism (Jaafar et al., 2013). In 2000, UNESCO added Kinabalu Park to its list of World Heritage Sites, which increased its visibility and appeal as a tourist attraction. This was the first UNESCO World Heritage Site in Malaysia. The primary stakeholders associated with Kinabalu Park are the Ministry of Tourism and Environment, the Wildlife Department, the Environmental Protection Department, the Sabah Cultural Board, the Sabah Tourism Board, residents, and tourists. Ensuring accessibility is one of their shared responsibilities.

4.1 Accessibility Study

Data is essential for documenting progress and for identifying areas for improvement. This gives us a sense of where we are and where we need to go. Therefore, we conducted a study to assess the accessibility and inclusiveness of these sites. As Parker (2001: 105) stated, “The level of accessibility is, to some extent, a physical manifestation of society's acceptance and respect for persons with different ability sets."

One of the authors, born in Malaysia, conducted the on-site observations, is fluent in Malay, English, and Mandarin, holds a bachelor's degree, and has research training and experience. Institutional Review Board approval was obtained. This study was conducted in 2023.

The researcher also looked for evidence of greater inclusiveness, such as the availability of alternative formats such as braille, tactile pavement, audio options for tours, and the use of sound in elevators and at traffic crossings. She took some photos and videos, as permitted, as additional documentation. She planned her visits, took notes on-site, and used local transportation. These observations and impressions added another dimension to this study. We incorporated this information in the following discussion of Kinabalu Park.

4.2 *Kinabalu Park in Sabah*

Kinabalu Park is located on the island of Borneo in the Malaysian state of Sabah. Kinabalu Park was formed over 15 million years ago by the granite intrusion of tectonic movements that shaped the land. It was established as a national park in 1964 after Professor Corner conducted a scientific expedition from 1962 to 1964. The park is known for its amazing diversity of plants, fauna, animals, and landscapes, such as rainforests and mountains. The park has the highest mountain peak in Southeast Asia, Mount Kinabalu (Sabah Parks, 2024). Additionally, Mount Kinabalu holds significant cultural meaning to the indigenous Kadazan people. It is believed that the spirits of their ancestors live on the mountain. The mountain's name comes from the term Aki Nabalu, meaning Revered Place of the Dead (Raikar, 2023). Kinabalu National Park became a UNESCO World Heritage site in 2000 to balance conservation with sustainable tourism. Infrastructures such as trails and education centers have been established to help facilitate visitors' access to the park while reducing the environmental impact. The Laban Rata is a rest house on Mount Kinabalu that provides overnight lodging for hikers. Guides, park rangers, and visitor centers help contribute to conservation efforts by increasing awareness of the park's importance and role in protecting the unique ecosystems for the future (Sabah Parks, 2024).

The researcher and ten hiker friends, none of whom had a disability, visited Kinabalu Park during April 5–7, 2023. Reservations are required since only 163 climbers per day are allowed. On April 5, we took a 2-hour 45-minute flight from Kuala Lumpur International Airport 2 to Kota Kinabalu International Airport. After that, we had lunch and traveled to the accommodation called Mile 36 Lodge, 1500 m elevation, via a rented private van. This journey took about 2 hours, and the paved asphalt road was rugged because it was narrow. In the Mile 36 Lodge, 11 of us stayed in the Willies Cottage, which has a maximum capacity of 12 people per night. There were no rooms designed for accessibility. The following day, we departed for the Kinabalu Park around 7:30 a.m. Kinabalu Park in Kundasang was about a 15-minute van journey from the Mile 36 Lodge (Fig. 4.2).

When we arrived, we registered ourselves and obtained our nametag for the hike at the visitor centre, which was accessible by everyone with a ramp slope at the side. There were different facilities available at Kinabalu Park such as Mount Kinabalu Botanical Garden, Kinabalu Natural History Gallery, Orchid Conservation Centre, Canopy Walk, Butterfly Farm, Poring Visitor Centre Gallery, and Rock Pool. Khew (2016) stated that not all services and facilities available in the park are accessible to people who use wheelchairs. Some may be accessible, partially accessible, or inaccessible. The park is partially accessible in terms of transportation, too. Visitors usually use private cars, hire taxis, or private vans to travel to the park. None of the

Fig. 4.2 The Willies Cottage before departing for the Kinabalu Park. (Photo by author on April 6, 2023)

public options is specially designed for people who use wheelchairs. At the same time, the accessible paved asphalt road in the Kinabalu Park Headquarters had a steep slope, which made it very challenging for people with disabilities, especially those with wheelchairs, to move around (Table 4.1).

Kinabalu Park contributes to the inclusiveness of people with disabilities by waiving the entrance fee, also called the conservation fee, for Malaysians in the OKU category (Orang Kurang Upaya, "People with Disabilities") while limiting inclusiveness with its fee structure. People who cannot afford the fee are prevented from visiting this national treasure, arguably owned by all Malaysians. On a more positive note, the park has allowed people with disabilities to climb Mount Kinabalu (4095 m), regarded as one of the highest peaks in Southeast Asia, with specific fitness preparation by the hikers. Although climbing Mount Kinabalu does not require any special skills or equipment to reach the peak, and it seems to be doable by anyone with a reasonable fitness level (Adventoro, 2018), mountain climbing can be an extreme activity for many people with disabilities.

Organized by the International Islamic University Malaysia (IIUM) in conjunction with the university's 40th anniversary, inspired by former IIUM student, Azzrulkhan Hasrullah Matsah, a disabled person who had paralysis in both lower limbs and successfully climbed Mount Kinabalu in 2016 and Everest Base Camp in 2018 (Thesun.my, 2022a), the Kinabalu OKU Challenge 2022 Expedition aimed to raise funds to enable IIUM's campus grounds to gain accessibility for the PWD community. Six out of the 22 participants were people with disabilities, and they reached the Mount Kinabalu peak on 31 August 2022. Among the six participants, OKU Sentral sustainability officer Khairull Hafiz Ismail was without a left leg while IIUM Economics and Management Science Associate Professor Mohd Nizam Barom (Doksil, 2022) and IIUM student Ahmad Danial Mohd Sani were visually impaired (Thesun.my, 2022b). This challenge promotes the abilities rather than the disabilities of people with disabilities and creates awareness that they are equally capable with the proper support. Their ability to conquer the highest summit in Malaysia boosts self-esteem and self-confidence, which are advantageous for success in life.

People who use wheelchairs and the visually impaired are advised to avoid visiting Kinabalu Park due to mobility issues with staircases everywhere. Other limitations are no accessible toilets and no pets or service animals permitted as they could

Table 4.1 Kinabalu Park entrance/conservation fees 2022–2024

Category	2022	2023	2024
Malaysians 18 years to age 59	3 MYR	10 MYR	10 MYR
Malaysians age 17 and younger	1 MYR	5 MYR	5 MYR
Malaysians below age 12, above age 60, and OKU (visitors with disabilities)	Free of charge	Free of charge	Free of charge
International 18 years old and older	15 MYR	50 MYR	50 MYR
International age 17 and younger	10 MYR	25 MYR	25 MYR

Source: By author based on Sabah Parks website

harm the habitat. However, anyone can access useful information about the park online. For example, the Kinabalu UNESCO Global Geopark Video has a short introduction to its geological, biological, and cultural attractions (Kinabalu Geopark, 2024). It portrays the breathtaking view of Mount Kinabalu (4095 m) and explains how it was formed to be the highest peak in Malaysia.

Sabah Parks, which administers Kinabalu Park, always welcomes researchers. They can apply for Sabah Parks Research ePermits (Jones, n.d.). As stated in the Sabah Park Strategic Plan 2025, the Board of Trustees of Sabah Parks would like Sabah Parks to be the platform for cutting-edge ecosystem research that is directed towards protected area management in tropical areas. In addition to their own researchers, they wish the Parks would be honored to have top-notch scientists and research programs from all corners of the world conduct research in the Parks (Lembaga Pemegang Amanah Taman-Taman Sabah, 2014:15). Research could include studying ways to increase inclusiveness and accessibility while preserving the ecology and culture of the area. This type of research can have a global impact as protected area managers grapple with over-tourism in our increasingly mobile and overpopulated world with eight billion inhabitants.

Research at Sabah Parks could generate fundamental knowledge about park management and many other aspects of tourism (Lembaga Pemegang Amanah Taman-Taman Sabah, 2014). Integration of research results and the management's decisions will aid in guaranteeing the long-term conservation of the park and its unique and fundamental natural values, as well as the inclusivity and accessibility of people with disabilities.

5 Conclusion and Recommendations

Sources from around the globe provide ample guidelines, regulations, research, practical experience, and tools for addressing inclusion, accessibility, and universal design. Applying this knowledge to the tourism sector remains a challenge. Although some progress is noted, to use Robert Frost's words "But I have promises to keep, and miles to go before I sleep, and miles to go before I sleep" (Frost, 1969: 225).

Based on our study, Kinabalu Park has room for improvement in the inclusiveness of PWDs by making their facilities more accessible while minimizing detrimental impacts on the quality of Kinabalu Park's environment. Zainol (2014) stated that nature-based tourism offers valuable experiences to visitors, which can only be appreciated or stimulated using sensory elements such as touch, taste, smell, sound, visual, and mobility. The more a visitor engages with their environment, the more senses are stimulated, which leads to a more memorable experience. Expanding access to the facilities in Kinabalu Park should enable higher satisfaction and greater inclusivity of the visitors including those with disabilities. Adding more virtual content on their website such as videos of the park and audio tours could be helpful.

Aljohani and Choy (2021: 123) generated many recommendations from their study of Kinabalu Park. These include improving infrastructure and public transportation and adding more festivals and other tourist-attracting activities. They advise careful implementation "to ensure that increased tourism does not destroy the qualities that attracted them." Making partially accessible features fully accessible, such as widening a trail to accommodate wheelchairs and installing an accessible toilet could be a start. Cantarellas (2023) developed the Heritage Site Accessibility Tool (HSAT) for archeological sites as an objective way to identify weaknesses. This tool could provide data to direct implementation efforts.

Well-managed and funded nature-based tourism can generate income for local communities, reduce poverty, preserve local culture, and incentivize residents to protect the natural environment (Gupta et al., 2023). Underfunding undermines effective management and threatens biodiversity and this remains a global concern (Deutz et al., 2020; Waldron et al., 2017). Increasing budgets for protected areas like Kinabalu Park combined with advertising campaigns targeting the economic and quality of life benefits for Malaysia is warranted.

Other recommendations include educating tourism personnel and the general public about universal standards and national regulations that are currently in place, and the need for expanding inclusion and accessibility. Another recommendation is more research to assess understanding and utilization and to ascertain tourism practitioners' views and aspirations gained from their experience in making tourism for people with disabilities a reality.

Addressing accessibility and broader inclusion also has workforce implications since persons with disabilities will have more employment opportunities in the tourism sector. This aligns with international goals, such as the U.N.'s 2030 Agenda for Sustainable Development, local goals, and applicable regulations. Another benefit is that greater inclusion can reduce inequality and marginalization of minority groups (Hon & Gamor, 2021).

Kinabalu Park is a classic example of the tension between disability rights codified in national regulations and nature-based tourism. Perhaps artificial intelligence (AI) can generate some solutions or create an in-person experience without requiring physical presence (Kosmas et al., 2019). The Dive into Heritage tool that UNESCO will launch in 2025 to facilitate remote exploration of World Heritage Sites is an example of possibilities (whc.unesco.org). Applying AI to UL could be a significant advancement for everyone which is the ultimate goal. We look forward to a day when all cultural heritage sites are accessible and inclusive.

References

Abdullah, N. B., Hanafi, H. B., & Hamdi, N. I. B. M. (2017). The rights of persons with disabilities in Malaysia: The underlying reasons for ineffectiveness of persons with disabilities act 2008. *International Journal for Studies on Children, Women, Elderly And Disabled, 1*, 127–134.

Adventoro. (2018). *Things to do in Kinabalu Park.* https://www.adventoro.com/blogs/adventure-news/things-to-do-in-kinabalu-park

Aljohani, N. M., & Choy, E. A. (2021). Impact of heritage tourism on local community in Sabah, Malaysia. *Journal of Social Sciences and Humanities, 18*(5), 122–137.

Arnett, J. J. (2002). The psychology of globalization. *American Psychologist, 57*(10), 774–783.

Austin, R. D., & Pisano, G. P. (2017, May–June). Neurodiversity as a competitive advantage. *Harvard Business Review, 95*, 96–103.

Balasubramanian, K., & Hanafiah, M. H. (2022). What innovations would enable the tourism and hospitality industry in Malaysia to rebuild? Worldwide Hospitality and Tourism Themes. https://doi.org/10.1108/WHATT-05-2022

Burgdorf, M. P., & Burgdorf, R., Jr. (1975). A history of unequal treatment: The qualifications of handicapped persons as a "suspect class" under the equal protection clause. *Santa Clara Lawyer, 15*, 855–910.

Cantarellas, M. A. S. (2023). Archeological heritage: A heritage site accessibility tool (HSAT) for open-air archeological sites. *Archeologies: Journal of the World Archeological Congress.* https://doi.org/10.1007/s11759-023-09487-0

CRPD. (2006). Convention on the rights of persons with disabilities: Optional protocol. AR/RES/61/106.

CRPD. (2014). *Committee on the rights of persons with disabilities.* Article 19.

Deutz, A., Heal, G. M., Niu, R., Swanson, E., Townshend, T., Zhu, L., et al. (2020). *Financing nature: Closing the global biodiversity financing gap.* The Paulson Institute, The Nature Conservancy, and the Cornell Atkinson Center for Sustainability https://www.paulsoninstitute.org/wp-content/uploads/2020/09/FINANCING-NATURE_Full-Report_Final-Version_091520.pdf

Ding, J. (2010, February 19). A disabled-friendly Malaysia? The nut graph. A disabled-friendly Malaysia? – *The Nut Graph.*

Doistua, J., Lazcano, I., & Madariaga, A. (2020). Self-managed leisure, satisfaction, and benefits perceived by disabled youth in northern Spain. *Frontiers in Psychology, 11*, 716. https://doi.org/10.3389/fpsyg.2020.00716

Doksil, M. (2022). Six persons with disabilities among 22 conquering Mt Kinabalu. *Borneo Post.* https://www.theborneopost.com/2022/08/30/six-persons-with-disabilities-among-22-conquering-mt-kinabalu/

Frances, L., & Silvers, A. (2016). Perspectives on the meaning of "disability.". *AMA Journal of Ethics, 18*(10), 1025–1033. https://doi.org/10.1001/journalofethics.2016.18.10.pfor2-1610

Frost, R. (1969). Stopping by woods on a snowy evening. In E. C. Lathem (Ed.), *The poetry of Robert Frost: The collected poems.* Henry Holt and Co.

Genç, R., & Genç, E. A. (2017). Promotion of social inclusion through new steps in tourism. *AUDŒ, 13*(3), 194–201.

Gupta, A, Zhu, H., Bhammar, H., Earley, E., Filipski, M., Nahrain, U. et al. (2023). Economic impact of nature-based tourism. *PLOS ONE* Dated April 12. https://doi.org/10.1371/journal.pone.0282912

Harniss, M. (2014). What is accessibility? In A. C. Michalos (Ed.), *Encyclopedia of quality of life and well-being research* (pp. 5–8). Springer. https://doi.org/10.1007/978-94-007-0753-5_8

Heine, S. J. (2020). *Cultural psychology* (4th ed.). W. W. Norton &.

Hon, A. H. Y., & Gamor, E. (2021). The inclusion of minority groups in tourism workforce: Proposition of an impression management framework through the lens of corporate social responsibility. *International Journal of Tourism Research, 24*, 216–226. https://doi.org/10.1002/jtr.2495

Iezzoni, L. I. (2016). Stigma and persons with disabilities. In R. Parekh & E. W. Childs (Eds.), *Stigma and prejudice: Touchstones in understanding diversity in healthcare.* Humana Press.

International Organization for Standardization. (2021). ISO 21902: 2021 Tourism and related services: Accessible tourism for all: Requirements and recommendations.

Jaafar, M., Kayat, K., Tangit, T. M., & Yacob, M. F. (2013). Nature-based rural tourism and its economic benefits: A case study of Kinabalu National Park. *Worldwide Hospitality and Tourism Themes, 5*(4), 342–352.

Jamaludin, M., & Kadir, S. A. (2012). Accessibility in buildings of tourist attraction: A case studies comparison. *Procedia-Social and Behavioral Sciences, 35*, 97–104.

Jepson, A., Stadler, R., & Garrod, B. (2023). Tourism and neurodiversity: A problematisation and research agenda. *Current Issues in Tourism*. Dated July 13. https://doi.org/10.1080/13683500.2023.2233040

Jones, D. (n.d.). Sabah Parks Research ePermit. https://researchpermit.sabahparks.org.my/

Kamarudin, H., Hashim, E., Mahmood, M., Ariff, N. R. M., & Ismail, W. Z. W. (2012). The implementation of the Malaysian standard code of practice on access for disabled persons by local authority. *Procedia—Social and Behavioral Sciences, 50*, 442–451.

Kamarudin, H., Ariff, N. R. M., Ismail, W. Z. W., & Ismail, E. W. (2013). Auditing Malaysian standard compliance for access and facilities for persons with disabilities in local authority buildings and its surrounding. In *3rd international conference on universal design in the built environment* (pp. 1–13).

Khew, E. H. (2016). *"Can the people who use wheelchairs enjoy the national parks?" compare the accessibility of people who use wheelchairs to national parks in Malaysia to Hokkaido's and create a spatial database for the future*. Master's Thesis, Rakuno Gakuen University, Japan. https://rakuno.repo.nii.ac.ip/record/3735/files/Khew.syuuron.pdf

Kim, M., Jasper, A. D., Lee, J., & Won, H. (2021). Work, leisure, and life satisfaction for employees with physical disabilities in South Korea. *Applied Research in Quality of Life, 17*, 469–487.

Kinabalu Geopark. (2024). *Kinabalu UNESCO global geoparks*. https://kinabalugeopark.sabahparks.org.my/

Kosmas, P., Galanakis, G., Constantinou, V., Drossis, G., Christofi, M., Klironomos, I., Zaphiris, P., Antona, M., & Stephanidis, C. (2019). Enhancing accessibility in cultural heritage environments: Considerations for social computing. *Universal Access in the Information Society, 19*, 471–482. https://doi.org/10.1007/s10209-019-00651-4

Laws of Malaysia. (2006). Act 645 National Heritage Act 2005. The Commissioner of Law Revision, Malaysia mal130377.pdf (fao.org)

Lembaga Pemegang Amanah Taman-Taman Sabah. (2014). Sabah Parks strategic plan 2025. Kota Kinabalu, Sabah.

Loo, K. K., Hutman, T., Yang, J. H., McAdam, D. B., & Nyp, S. S. (2021). Disorders, disabilities, and differences: Reconciling the medical model with a neurodiversity perspective. *Journal of Developmental & Behavioral Pediatrics, 42*(9), 763–766. https://doi.org/10.1097/DBP.0000000000001032

Majdi, N. B. (2021). *Accessibility of disabled toilet for wheelchair user in public buildings in Kuala Lumpur*. Master's Thesis, University of Malaya.

Malaysia: Federal Constitution. (1957, 31 August). https://www.refworld.org/legal/legislation/natlegbod/1957/en/40703

Malijani, O. (2019). Towards community inclusion: Chongoni world heritage site and and sustainable cultural tourism in Malawi. *Southeast African Archaeological Bulletin, 74*(209), 35–41.

Marsin, J. M., Arrifin, S. I., & Shahminan, R. N. R. (2014). Comparison of legislation concerning people with disability and heritage environment in Malaysia and developed countries. *Earth and Environmental Science, 18*, 0121278. https://doi.org/10.1088/1755-1315/18/1/012178

Martin, L., & Cobigo, V. (2011). Definitions matter in understanding social inclusion. *Journal of Policy and Practice in Intellectual Disabilities, 8*(4), 276–282.

Murray, D. R., & Schaller, M. (2016). The behavioral immune system: Implications for social cognition, social interaction, and social influence. *Advances in Experimental Social Psychology, 53*, 75–129.

Othering & Belonging Institute. (2020). Inclusiveness index: Measuring global inclusion and marginality. *Global Inclusion Index 2020*. https://escholarship.org/uc/item/4vw5c7z0

Parker, K. J. (2001). Changing attitudes towards persons with disabilities in Asia. *Disability Studies Quarterly, 21*(4), 105–113.

Rahim, A. A. (2021). Paper 1B. MS 1184:2014 universal design and accessibility in the built environment – Code of practice https://architecturemalaysia.com/Files/Pool/177_211110_1659555955_paper_1_06112021_pam_northern_chapter_webinar_ms1184pptx.pdf

Raikar, S. P. (2023). Mount Kinabalu | Geography, Mythology, Trivia, & Facts. Britannica. Dated November 6. https://www.britannica.com/place/Mount-Kinabalu

Sabah Parks. (2024). *Mount Kinabalu climbing*. https://www.sabahparks.org.my/services/mount-kinabalu-climbing

Saravanabhavan, S., & Saravanabhavan, R. C. (2001). Attitudes towards disabilities across cultures. *Educational Practice and Theory, 23*(2), 49–60.

Scheyvens, R., & Biddulph, R. (2017). Inclusive tourism development. *Tourism Development, 20*, 1–21. https://doi.org/10.1080/14616686.2017.1381985

Sirgy, M. A., Uysal, M., & Kruger, S. (2017). Towards a benefits theory of leisure well-being. *Applied Research in Quality of Life, 12*, 205–228. https://doi.org/10.1007/s.1148-016-948-7

Smallman, S., & Brown, K. (2020). *Introduction to international and global studies* (3rd ed.). The University of North Carolina Press.

Thesun.my. (2022a). *22climbersinOKUchallenge2022settoconquerKinabaluSummit*.https://thesun.my/home/22-climbers-in-oku-challenge-2022-set-to-conquer-kinabalu-summit-MJ9759853

Thesun.my. (2022b). *Kinabalu Oku Challenge 2022 participants reach top of Mount Kinabalu*. https://thesun.my/home/kinabalu-oku-challenge-2022-participants-reach-top-of-mount-kinabalu-ED9764071

U.S. Department of Labor. (1990) Americans with Disabilities Act, 42 USC sec 10102(1) https://www.dol.gov/general/topic/disability/ada

UN-ESCAP. (1993). (United Nations Economic and Social Commission for Asia and the Pacific) Asian and Pacific decade of disabled persons, 1993–2002: The starting point. https://hdl.handle.net/20.500.12870/4381

UNESCO. (2023). *Kinabalu Park*. World Heritage Convention. https://whc.unesco.org/en/list/1012/

United Nations Educational, Scientific and Cultural Organization (UNESCO) World Heritage Convention (n.d.). Dive into heritage. https://whc.unesco.org/en/dive-into-heritage/.

UNESCO. (n.d.). *World heritage list*. World Heritage Centre UNESCO World Heritage Centre – World Heritage List.

United Nations. (2006). Convention on the rights of persons with disabilities. *Treaty Series, 25153*, 3.

United Nations. (2023). Transforming our world: The 2030 agenda for sustainable development. https://sdgs.un.org/2030genda

United Nations. (n.d.). Social inclusion. Department of Economic and Social Affairs. Convention on the Rights of Persons with Disabilities (CRPD) | Division for Inclusive Social Development (DISD) (un.org).

UNWTO. (2023). (United Nations World Tourism Organization) accessible tourism. https://www.unwto.org/accessibility

Waldron, A., Miller, D. C., Redding, D., Mooers, A., Kuhn, T. S., Nibbelink, N., et al. (2017). Reductions in global biodiversity loss predicted from conservation spending. *Nature, 551*(7680), 364–367.

Wikimedia Foundation. (2023). List of world heritage sites in Malaysia. https://en.m.wikipedia.org/wiki/List_of_World_Heritage_Sites_in_Malaysia.

World Health Organization (WHO). (2017). *International classification of functioning, disability, and health social activities as a fundamental right*. United Nations.

World Health Organization (WHO). (2021). *International classification of functioning, disability, and health*. United Nations.

World Health Organization (WHO). (2023). *Disability: Key facts*. https://www.who.int/news-room/fact- sheets/detail/disability-and-health

Zahari, N. F., Che-Ani, A. I., Abdul Rashid, R. B., Tahir, M. A. M., & Amat, S. (2020). Factors contribute in development of the assessment framework for wheelchair accessibility in National Heritage Buildings in Malaysia. *International Journal of Building Pathology and Adaptation, 38*(2), 311–328. https://doi.org/10.1108/IJBPA-02-2019-0021

Zainol, R. (2014). Appreciating a world heritage site using multisensory elements: A case study in Kinabalu Park, Sabah. *Malaysia. SHS Web of Conferences, 12*, 01080. https://doi.org/10.1051/shsconf/20141201080

Meg Milligan, PhD in Sociology and Anthropology, American University of Beirut, Lebanon, PhD in Counseling Psychology, Auburn University, Alabama, USA, psychology internship at the Veterans Administration Medical Center, Tuskegee, Alabama, licensed psychologist in Alabama, is a Professor in the Department of Psychology and adjunct faculty in the global leadership doctoral program at Troy University, Alabama. Milligan has thirty years of teaching experience in higher education in Malaysia (2016–2022) and the United States, conducts study abroad programs annually, has presented over 100 papers and posters at conferences in nine countries spanning three continents, has faculty, national research, and international research excellence awards, and publications on disability rights, narcissistic leadership, and global identity among others. Milligan is President-Elect (2023–2024) of the Association for Distance Education and Independent Learning (ADEIL) after two terms on their Board of Directors, will be President (2024–2025), has Quality Matters (QM©) master reviewer certification at all levels.

Lee Ting Lew holds a BA in Psychology from SEGi College Kuala Lumpur, Malaysia, attained through the American Degree Program in collaboration with Troy University, Alabama, USA. Graduating in 2020, she brings three years of valuable experience in Human Resources to her endeavors. Beyond the professional realm, she finds solace and inspiration in nature, enjoying activities such as travelling and hiking. Firmly believing in the therapeutic power of the natural world, she advocates for its integration into daily life as the ultimate remedy for humanity. Passionate about giving back to the community, she actively engages in charity and volunteering initiatives, acknowledging the significance of contributing to societal well-being. Embracing diversity and different cultural perspectives, she is a member of the American Field Service Malaysia and Rescue Union Malaysia, promoting inclusivity and mutual understanding in all her endeavors, thereby enriching experiences and fostering meaningful connections across boundaries.

Suzanne Tew-Washburn has a BA in Sociology from Columbus College, Georgia, an MS degree in personnel from Troy University, USA and a PhD in rehabilitation counseling from Auburn University, Alabama. She is now Associate Professor at Troy University, Director of Rehabilitation Counseling Education on nine Troy campuses, Project Director of rehabilitation counseling scholarship grants totaling three million dollars, and Director of Operations for the College of Education on her home campus She secured and managed over 12 million dollars in grants, was awarded Troy University's Chancellor's Award of Distinction for Sponsored Program Success (2010, 2015), and received the Rehabilitation Services Administration Commissioner's Award for Excellence in Education and Training from the US Department of Education. Before joining Troy University, she was an educator and distance learning director at Auburn University and a practitioner at a community rehabilitation agency in Georgia. Her publications and presentations include job development training, rehabilitation employment services, accreditation and credentialing, and accessibility and inclusion.

Mary Kristin Medina graduated from Troy University, Alabama, USA with a BS degree in Psychology and an MS degree in Clinical Mental Health Counseling. Ms. Medina served as the graduate assistant for her program at Troy University, where she assisted in research about play therapy. Ms. Medina is a member of the Chi Sigma Iota honor society. In her free time, she enjoys volunteering at her local animal shelter. Ms. Medina aspires to work with children and their families while continuing to pursue her interests in various research areas and further her studies in a PhD program.

Chapter 5
Folklore for Cultural Tourism: The Malaysian Experience

Hoon-Peow See

1 Introduction

When Kampung (Village) Nalapak, a community by the Moroli River near Kundasang in Sabah, East Malaysia was opened to the public in 2021, it instantly attracted many tourists who looked for an alternative experience from the local routine sites in the Ranau highlands. The main attraction of Kampung Nalapak, is a split rock, which received its name from Village Nalapak, a term meaning 'split' in the Kadazandusun language. What makes the rock fascinating is the folklore associated with it which has been passed down through generations. According to the folklore, the rock was split by a mythical creature called 'Tumotolong', a mother with a profound love for her child. A malevolent crocodile took her child away and hid the child in a rock. Out of profound love for her child, she sought for her child relentlessly and found the child in the rock. She had no choice but to split the rock to release the child. The fascinating story gave the split rock an additional attraction, other than its natural beauty. With great wisdom the Sabah Tourism, Culture, and Environment Ministry recognized the significance of folklore, particularly this captivating tale in enhancing the value and allure of rural community-based tourism. In the words of Datuk Joniston Bangkuai, the Assistant Tourism Minister, he said:

> The impact of storytelling extends beyond the realm of entertainment. It provides a livelihood for the community, empowering them to take charge of their narrative. When storytelling and rural community-based tourism come together, a transformative experience takes shape. Visitors leave with fond memories and a profound understanding on the importance of keeping different cultures alive. They become ambassadors, sharing the tales they heard and spreading the word about the remarkable community they encountered.
>
> It is through these profound narratives that a simple village becomes a vibrant tapestry of traditions that invites the world to listen, learn, and be inspired (Geraldine, 2023).

H.-P. See (✉)
Sunway TES College, Iskandar Puteri, Johor, Malaysia

T.-C. Wong et al. (eds.), *Cultural Tourism in the Asia Pacific*,
https://doi.org/10.1007/978-3-031-63459-8_5

The mission of this chapter is to explore the value of folklore, especially folk narratives in attracting and providing rich tour experiences to visitors.

2 Definition and Genres of Folklore

Folklore is not easy to define but it is easy to identify when encountered. Since the British scholar, William Thomas coined the term 'folklore' to replace 'Popular Antiquities' and 'Popular Literature' in 1846, it has been defined in many ways (Dundes, 1965: 4–6). Literally, the term consists of two parts: 'folk' and 'lore'. Some of the definitions emphasize the 'folk' while the other focus on the 'lore'.

One of the most influential definitions of folklore that is accepted by many folklorists today is that of Dundes (1965, 3) which simply outlines a non-exhaustive list of genres of folklore as follows:

> Folklore includes myths, legends, folktales, jokes, proverbs, riddles, chants, charms, blessings, curse, oaths, insults, retorts, taunts teases, toasts, tongue-twisters, and greeting and leave-taking formulas (e.g., See you later, alligator). It also includes folk costume, folk dance, folk drama (and mime), folk art, folk beliefs (or superstition), folk medicine, folk instrumental music (e.g., fiddle tunes), folk songs (e.g., lullabies, ballads), folk speech (e.g., slang), folk similes (e.g., blind as a bat), folk metaphors (e.g., to paint the town red), and names (e.g., nicknames and place names)…

The most important criterion in defining folklore is that folklore is supposed to be transmitted orally (Dundes, 1965: 1). Elsewhere Dundes has also emphasized on the importance of anonymity and multiple versions in identifying folklore. Due to the unknown origin (author), multiple versions existed due to the long history of transmission from generation to generation orally (usually); yet, interestingly the main storyline is stable (Dundes, 1975: 3–14).

However, not all stories of folklore are transmitted orally (Dundes, 1965: 25). There are some genres of folklore such as autograph-book verse, book marginalia, epitaphs and traditional letters, which are transmitted in written and not in oral form. Some folklore such as folk dance, games and gestures may be arguably transmitted through observation rather than orally (Dundes, 1965: 1–2). Thomson (1946: 5) also points out that "it is impossible to make a complete separation of the written and oral traditions". This is due mainly to two reasons: Firstly, some tales may be collected and written down. Secondly, written stories may be passed on by word of mouth and become oral stories. In this chapter, we shall mainly discuss the relevance of folk narratives and performances to tourism.

The term 'folk' often leads one to think of simple rural villagers, like in the story shared in the Introduction. However, the term in fact refers to "any group of people whatsoever who share at least one common factor" (Dundes, 1965: 2). The common factor can be anything, as long as it is enough to form a group with a certain tradition on its own. This tradition is very important as it forms a certain identity of the folks. Losing this identity may mean losing their psychological equanimity. An identity is an important human psychological need. Losing it will then make the folk feel insecure and restless. Thus, identity education is a very important aspect of socialization. This point is of significant importance as further discussions ensue.

In this study, I generally assume that folk narrative includes: myths, legends, folktales, epics, folk dramas etc. Myths are in general accepted as stories about extraordinary explanations of natural phenomena. The most commonly discussed myths are about the creation of the universe, the origin of mankind, the flood, astronomy and cultural heroes.

While the emphasis of myths is the spirituality of the characters, the emphasis of legends is the humanity of the characters. Myths are stories about gods; legends are stories about humans, mainly historical heroes. They are usually improvised stories of historical events, stories about the origin of a place, custom or cultural artifacts. However, interestingly, the main characters of some legends are mythical characters such as the mythical Foxes (Werner, 1994: 370–385). Perhaps the difference is in the emphasis of these stories – myths emphasize the spirituality of these characters and legends emphasize the humanity of these characters. Chinese scholars have indeed long observed that a myth can change into a legend, thus making it quite difficult to strictly differentiate between myths and legends.

Folktales are short stories with some consistent features, which we will explore more thoroughly later. The Aarne-Thompson index system (AT System) categorizes them into seven broad groups (Thomson, 1946: 481–487).

Index List	Groups
1–299	Animal Tales
300–749	Tales of Magic
750–849	Religious Tales
850–999	Realistic Tales
1000–1199	Tales of the Stupid Ogre (Giant, Devil)
1200–1999	Anecdotes and Jokes
2000–2399	Formula Tales

Myths and Legends are also considered folktales in indexing system, though they possess distinctive characteristics. Thus, the term 'folktale' has a broad sense which includes myths and legends and a narrower sense which excludes them.

Epics are poetic narratives which have characters, plots and are narrated from a third person's perspective. They can be classified into three categories, (a) epics about creation; (b) epics about heroes; and (c) epics about love stories. The epics about creation can also be considered as myths.

However, it is advisable to remember Thomson's (1946: 7) warning: "With the folktale, as with all other products of man's artistic endeavour, the scholar runs the risk of too subtle analysis. He may interest himself in studying the entire body of oral narrative of a person so as to divide it neatly into categories, according to origin or form or content, and although such a close examination of the stories undoubtedly teaches him much, he must realize that the men and women who told them, neither know nor care about his distinctions. Much hair-splitting has taken place in the past and much useless efforts devoted to the establishment of exact terms for the various kinds of folktale".

Thomson himself is very liberal in the use of the term 'folktale' to include many types of folk narrative. Therefore, hair-splitting effort should be generally avoided. It is important to bear in mind these definitions and genres of folklore as they help us in our analysis on its use of 'folklore' in tourism later.

3 Cultural Tourism and Functions of Folklore

Cultural tourism, in its broadest sense, encompasses a diverse range of activities. While its precise definition remains a subject of ongoing debate, one constant element within any definition is the core objective of gaining an understanding of a particular culture (Richards, 2003). Folklore, as a vital aspect of culture, unquestionably holds significance in the context of cultural tourism.

Folklore, in particular folk narratives, plays a number of functions in non-literate societies – e.g., entertainment, validation of culture, transmission of values, and the like. However, the functions of folklore may depend on the social context (Bascom, 1965). One of the major functions of folklore in non-literate societies is education. This has been well researched and documented in many parts of the world. However, it should also be realized that the relationship between functions and any particular item of folklore also varies according to social context, and that these functions may be interrelated. According to Dundes (1965: 279):

> Different types of folklore can share similar, if not identical, functions … The function is the same; the form is different. It is equally important to realize that any one item of folklore may have several different functions…The form is the same; the function is different. One cannot always tell from form alone what the associated contextual function is. Functional data must, therefore, be recorded when the item is collected. An item once removed from its social context and published in this way deprives the scientific folklorist of an opportunity to understand why the particular item was used in a particular situation to meet a particular need.

While the educational function of folklore in non-literate societies is very well documented, the value of folklore to a modern literate society is very much unexplored. In fact, Penjore's (2005: 64) comment on the Bhutan situation is relevant for many communities:

> We have desperately failed to use the strength and wisdom of the urban populace and promote them in schools through curriculum and to the larger population through mass media like TV, radio, press and film. This failure can be largely attributed to the government and public failure to recognize and appreciate values and wisdom represented by the people.

This comment provides neatly food for thought to the education of tourists about the local culture. In this chapter, we are going to explore the value of folklore in attracting tourists and educating them about the communities they have visited. This enhances their cultural experiences.

4 The Malaysian Experience

The Ministry of Tourism Malaysia had long realized the value of Malaysian folklore in attracting tourists. A number of myths and legends were included in the descriptions of popular tourist attraction spots in the promotional materials published since 1969 (Razak & Romle, 2016: 1380–1381). In particular, those associated with major attractions such as the Langkawi Island and Mount Kinabalu (Razak & Romle, 2016; Razak & Ibrahim, 2017).

As an example, a 47-page booklet, entitled 'Treasured Malaysian Legends', was published by the Ministry of Tourism Malaysia, with the major theme "Malaysia Truly Asia" in 2007 (Tourism Malaysia, 2007). One to a few myths or legends from each state of Malaysia and the Federal Territory of Labuan were included in the booklet. Again, the myths and legends selected were associated with major tourist attractions in each state.

Interestingly, for publicity purposes, the term 'myth' was dropped and all stories have been termed as legends, despite the fact that the booklet's introduction continues to refer them clearly as 'mystifying tales' and 'magical stories' as well as describing them as 'rich in adventure and sagas about fairies, heroes, magic, curse and heavenly celestial beings' (Tourism Malaysia, 2007: 2). Table 5.1 below gives a summary of the stories selected.

From Table 5.1 below, we can clearly see that Kedah had contributed far more myths and legends than any other states. Four out of six stories selected here are dedicated to the Langkawi Island. This may be due to the fact that Kedah has the longest history amongst the Malaysian states. This is an important point to consider in the later discussion.

All the stories selected are old Malay or indigenous stories, except for Sri Rambai—The Magic Cannon and The Giant Footprint from Penang; The Intriguing Tales of Melawati Hill and The Mysterious Jugral Lighthouse, from Selangor; which are associated with the colonial power in the past. However, the story about Sri Rambai is actually not folklore but factually historical. The only folkloristic thing about it is the belief that it is magical. The Chinese and Indian versions, alongside the Malay version, were presented in the story of The Mysterious Giant Footprint. While the Chinese were mentioned in the Kinabalu Dragon Gem story, all these stories dated back to at least the nineteenth century.

The Chinese character cited in the Mount Kinabalu legend is actually an historical figure who was known in three different names as Wong Song Ping, Wong Sum Ping and Ong Sum Ping. He arrival by ship to Brunei in around 1375, and upon his arrival at the Sungai Kinabatangan, his crew members broke his arms and according to the local folklore, this incident gave the place a name called 'Kina Patah Tangan' (an arm broken Chinese). Another more official appellation for the term Kinabatangan is *Sungai Cina* (Chinese River) which has indicated the Chinese were active in that region (Wong, 2011; Wikipedia, 2013).

Actually some of the stories are not legends or myths, they are actually folktales. For example, the stories about the mysterious Jugral Lighthouse, the Turtle Stones

Table 5.1 Treasured Malaysian Legends

State/ territory	The legends	Attractions
Kedah	**1. Lankawi's Legendary Brawl** The three islands of Langkawi are believed to have been transformed from three giants in time immorial. Giant Mat Raya and giant Mat Cincang fought over an engagement dispute and they were separated by another giant Mat Sawar in between. Tanjung Cincin nearby was the engagement ring and the village *Belanga Pecah* was a broken piece of crockery.	1. Legend Park; Kuah Jetty.
	2. The Seven Magical Wells According to the legend of "Telaga Tujuh," or the Seven Wells waterfall, it was considered the playground of fairies and sprites. There were seven sprightly fairies, each associated with one of the seven pools at the waterfall, believed to have healing properties. When the fairies visited, the air was filled with a delightful fragrance, and their giggles could be heard if one listened carefully.	2. Waterfalls
	3. Mahsuri…A Woman Wronged Mahsuri, born to a childless couple, was exceptionally beautiful and married Mat Deris, Langkawi's ruler's son. During Mat Deris's trip to Siam for trade, Mahsuri befriended a minstrel named Deramang. False accusations of adultery by her jealous mother-in-law led to her being sentenced to execution, by her father-in-law, who secretly desired her. Astonishingly, she remained unharmed by spears. But she revealed that only a specific sword could harm her. When she was stabbed, her blood was white. In her dying moments, she cursed Langkawi for seven generations. After her death, Langkawi experienced war with Siam and famine due to rice burning, enduring decades of misfortunes attributed to Mahsuri's curse.	3. Mahsuri's Tomb; Mahsuri's Well; Mahsuri Memorial and burnt rice grains at Padang Mat Sirat.
	4. The Magical Lake of The Pregnant Maiden Two stories are associated with the lake. The first involves Mat Teja, a male genie, and Mambang Sari, a beautiful female sprite. They were married, but their baby died shortly after birth, and in her grief, Mambang Sari laid the child to rest in the lake, imbuing it with magical powers. In the second story, a king's maid named Dayang Telani became magically pregnant after drinking from the lake during a hunting excursion. The king was angered and ordered her to live by the lake, where she gave birth to a son. Tragically, the boy drowned in the lake, and Dayang Telani begged the magical lake to take her in his place. As the boy drowned, a white crocodile emerged, and she was transformed into a rock known as Batu Dayang or Maiden Stone.	4. *Tasik Dayang Bunting*

	5. The Mystical Cave of Stories According to the Hikayat of Merong Maha Wangsa, the Garuda, a mythical bird, saw the union of a Chinese princess and a Roman prince as a threat to smaller kingdoms. To prevent their marriage, the Garuda kidnapped the princess and hid her in Gua Cerita, the Cave of Stories in Langkawi. He then attacked the Roman Prince's fleet, destroying it. The Roman Prince survived and washed ashore in Langkawi, where he met the princess, and they fell in love. In response, the Garuda, feeling disgraced, fulfilled a vow by setting himself on fire. Merong Mahawangsa, believed to have supernatural origins and in command of the prince's fleet, chose not to return to Rome, believing the princess had perished. Instead, he established the kingdom of Langkasuka.	5. *Gua Cerita*
	6. Raja Bersiong - Kedah Legendary Fanged King According to Hikayat Merong Mahawangsa, Raja Ong Maha Perita Doria of early Langkasuka developed an addiction to human blood after an accidental encounter. His incisors began to elongate into fangs, driving him to order the capture and killing of villagers for their blood, imprisoning them in Gua Penjara, the Prison Cave near today's Pantai Merdeka beach town. As the oppression continued, the villagers eventually rebelled against Raja Ong. He fled into the jungle, where a shaman transformed him into a wild boar.	6. Lembah Bujang Archeological Museum; Temple of the Ninth Water pool; Raja Bersiong's tomb in Kampung Seberang Tok Soh.
Penang	**1. Sri Rambai—The Magic Cannon** The cannon '*Sri Rambai* at Fort Cornwallis' is a historically significant piece believed to possess magical qualities. Originally a gift from the Dutch to the Sultan of Johor in 1606, it had a turbulent history. Seized by the Portuguese and taken to Jawa in 1613, it was later brought to Kuala Selangor in 1795. However, during an attempted shipment to Penang in 1871, it was stolen by pirates and thrown into the sea near the Esplanade. Multiple recovery attempts failed until its successful salvage in 1880. With its rich history of changing hands and enduring numerous adventures, the cannon stands as a unique and storied artifact.	1. Fort Cornwallis; *Sri Rambai.*
	2. The Mysterious Giant Footprint The legend of Big Foot in Penang revolves around a large footprint discovered in Batu Maung, a fishing village near the island's southern tip, measuring 33 inches in length. This enigmatic footprint has led to various interpretations among local communities. Indians believe it to be the footprint of the Indian God Hanuman, suggesting he took a colossal step while crossing the ocean. Malays attribute it to Gedembai, a giant, while the Chinese locals connect it with Admiral Cheng Ho and have even erected a temple called 'Sam Poh Footprint' to honor the legendary print. This site exemplifies how diverse cultural beliefs and legends can intersect in a single location.	2. *Batu Maung*

(continued)

Table 5.1 (continued)

State/ territory	The legends	Attractions
	3. The Adventures of the Singing Sea Captain In the past, the Malays were renowned seafarers, with mariners like Nakhoda Ragam, often called the “singing captain” due to his musical talents. Nakhoda Ragam played a significant role in naming Penang during his trading voyages in the Straits of Malacca. He discovered a large isolated island, naming it ‘Pulau ke Satu’ or ‘single island.’ His water-pot, or kendi, fell into the sea at the island’s southern coast, becoming Pulau Kendi. Releasing a tame bird, Bayan, at the southern coastline led to that area being known as Bayan Lepas. When he encountered a storm near Pulau Betong on the western side, his wife’s pale face gave the place the name Pucat Muka, meaning ‘pale faced.’ These stories and legends add a rich historical and cultural layer to Penang’s heritage.	3. Penang; Bayan Lepas; *Pucat Muka*
Perak	**1. Perak Royalty Rituals** The present Sultanate of Perak, established in 1528, traces its origins to Sultan Mudzaffar Shah, the eldest son of the last Sultan of Malacca, through a legendary tale. While en route to Perak, Sultan Mudzaffar Shah encountered a violent storm. An immense genie materialized and instructed him to throw his crown into the sea if he wanted to become the Sultan of Perak. Fearing for everyone’s safety, he followed the genie’s command. Upon reaching Beting Beras Basah, he proclaimed the establishment of his Sultanate.	*1. Muzium Diraja Kuala Kangsar*
Selangor	**1. The Intriguing Tales of Melawati Hill** During the reign of the second Sultan of Selangor, Sultan Ibrahim, a fort was constructed to defend against Dutch invasions. Numerous battles took place here, and a local legend tells of a girl in the fort who was beheaded on the Batu Hampur, the executioner’s stone block, for committing adultery. Her blood was spread around the fort as a warning against defying the law.	1. Melawati Hill; Fort Altingsburg; Altingsburg Lighthouse; *Batu Hampur*.
	2. The Mysterious Jugral Lighthouse A lighthouse was constructed on a hill known locally as ‘Bukit Kerang’ or Cockle Hill. The building of the lighthouse on this sacred peak was believed to have violated it, resulting in the hill bleeding continuously for 30 days. This caused the water around the hill to take on a rusty-red color resembling the cockles found in the area. Ironically, the cockle population began to decline and eventually vanished not long after this incident.	2. Jugra Lighthouse

Johor	**1. The Fury of Laksamana Bentan** Laksamana (Admiral) Bentan was a fearless warrior who served under Sultan Mahmud Shah II and was tasked with ending a rebellion in the territory of Haru and Jambi. The Sultan promised to care for Bentan's pregnant wife, Wan Anum, during his absence. However, a fateful incident occurred when Wan Anum ate a piece of jackfruit, from the first harvest before the Sultan, against the custom. Wan Anum repeatedly apologized, explaining that it was her child who needed the fruit when the Sultan discovered the her act. Wan Anum was killed when her stomach was cut open to ascertain that it was indeed the child who craved for the fruite. Upon hearing this devastating news, Laksamana Bentan was filled with fury. When he returned to Kota Tinggi, he unleashed havoc in the palace and killed the Sultan. Before the Sultan's death, he flung his kris (dagger) at Bentan's feet and cursed his next seven generations, foretelling that they would suffer from a condition causing them to emit blood and die should they ever come to Kota Tinggi. This tale is a tragic account of power, revenge, and the consequences of a vengeful act.	1. Sultan Mahmud II's tomb in *Kampung Makam; Makam Kucing Bertanduk;* Kota Tinggi Historical Centre; Kota Tinggi Waterfall; Laksamana Bentan's tomb in Kampung Kelantan.
	2. The Tragic Tales of Buaya Sangkut & Upeh Guling Buaya Sangkut and *Upeh Guling* are two captivating waterfalls in the Endau-Rompin State Park. Upeh Guling is named after a young Orang Asli, Upeh, who tragically slipped and rolled down the waterfall to his death while trying to cross the river when he saw his fiancée. Buaya Sangkut is a five-stepped waterfall with a tale involving a family trying to protect their baby girl from a crocodile. Despite their efforts, both the crocodile and the python summoned to fight it perished. The crocodile's skin was hung on the fall but fell on the girl, killing her. Many believe the crocodile-shaped stone at the fall's base is the same crocodile.	2. Endau-Rompin State Park
Melaka	**1. The Fairy Princess of Gunung Ledang (Mount Ophir)** Sultan Mansur Shah wished to marry the beautiful Fairy Princess of Gunung Ledang. The Princess's guardian, Dang Raya Rani, gave the Sultan the task to acquire seven trays of mosquito hearts; seven trays of hearts of mites; a vat of water from dried area nuts; a vat of tears from virgin maidens; a bridge of gold from Malacca to the peak of Gunung Ledang and a bowl of his young prince's blood as betrothal gifts to dissuade him. The cruel Sultan agreed to the dowry and oppressed his people to acquire them. Just as he was about to stab his own son for his blood, the Fairy Princess appeared. She swore would never marry such a cruel person. She eventually married Nakhoda Ragam, a renowned seafarer. However, she accidentally killed him with a needle. She returned to Gunung Ledang, heart-broken and swore not to set eyes on another man.	1. Gunung Legang This is probably the most famous legandary story from Malaysia. A movie was made in 2004, the princess was played by Tiara Jacquelina

(continued)

Table 5.1 (continued)

State/ territory	The legends	Attractions
Pahang	**1. The Dragon of Lake Chini** The Jakun people have a legend that speaks of an ancient and magnificent city, as grand as Angkor Wat, lying at the bottom of Lake Chini. However, the most celebrated legend of Lake Chini is about the Naga (Dragon) Seri Gumum. The Jakuns, who were skilled hunters and expert farmers, once encountered an old woman while clearing land for cultivation. She declared the area as hers and warned them not to infringe on it, by planting her walking stick in the ground. After apologizing, the Jakuns were allowed to continue their work. However, days later, their dog became agitated by a mossy-colored, log-like shape near the forbidden area. When one of the Jakuns threw his stick at it, the log started to move and bleed, causing a catastrophic event with darkened skies, lightning, and thunder. In their panic, a man knocked over the old woman's walking stick, and a fountain of water shot into the sky. This water continued to flow, forming the lake, which is now the dwelling place of Naga Seri Gumum, transformed from the magical log.	1. Lake Chini, 100 KM from Kuantan. Famoust for it refreshing lush of vegetation and a popular spot for bird watching.
	2. The Legend of Kota Gelanggi Cave Legend has it that the caves were actually the ruin of an ancient city of forest elves (Bunian), which were turned into stones together with their belongings. Others believed Kota Gelaggi was a hiding place of Princess Bera and Princess Jelai who eloped due to their parents' oppression. It resulted in a civil war in *Inderapura* (City of Demigods), an ancient kingdom of Pahang.	2. Kota Gelaggi Cave, 30 km east of Jerantut, Pahang. Not to be confused with the lost city of Kota Gelaggi in Johor. Kota Gelaggi Museum
	3. The Isle of the Sleeping Dragon In a local legend, the South China Sea was once home to a family of powerful dragons. A dragon prince and princess set out to explore the sea, but the princess lost her silk bridal shawl. As she went to retrieve it, her husband were not found upon her return. She waited for centuries, holding the shawl, and was eventually transformed into the beautiful Tioman Island with granite twin peaks, known as Chula Naga or Dragon's Horns, after a sprite cast a sleeping spell on her out of pity.	3. Tioman Island

Trengganu	**1. The Tragic Tale of A Turtle Stone** According to the story, a turtle couple was playing near a beach and they realized that they were lost. The male turtle decided to venture deeper onto the shore to see a beautiful garden that they were told. Unfortunately, he was lost and drank from a forbidden pond and turned into a stone. Thus, explains the shape of the stone. A seagull informed the female turtle about the bad news. The female returned to the sea grieving for her unfortunate companion. However, no matter where she went, she will always come back to the same place to lay eggs and shed tears thinking of him. This place is called Batu Penyu or Turtle Rock.	1. Rantau Abang; Turtle watching
	2. The Kind Princess at Princess Hill The hill got its name from different local legends. Some say it was named after a Bunian (Forest elf) Princess who used to live there. She used to lent her silver and gold plates and pots to locals for marriage feasts, but many didn't return or even broke them, causing her to move to Bukit Bintang. Gua Puteri was where she kept her precious items. Others claim it was named after seven benevolent elf princesses who resided there, the same elf princesses summoned in the haunting ancient melody, Ulek Mayang.	2. Bukit Puteri Font; *Gua Puteri; Seri Jamlor* and *Meriam Beranak* cannon.
	3. The Legend of Perhentian Island In ancient times, an immense and mighty dragon known as Naga Pulau Berjuang inhabited the island. One day, another male dragon attacked, and Naga Pulau Berjuang descended swiftly to engage in battle. His descent left a marked trail on the island, now recognized as Jalan Naga. According to another legend, the island remains free of poisonous snakes due to the protective power of the guardian dragon, as well as the presence of *keladi lekir* (Araceae Amorphophalus) and *akar cina putih* (white Chinese roots), which effectively neutralize poison.	3. Perhnetian Besar and Kecil; snorkelling; jungle trecking; beaches.
Kelantan	**1. The Legend of Princess Sa'adong** She was found as a baby in pure white bubbles, was raised by the legendary warrior Queen Che' Siti Wan Kembang in Kelantan. She later married her cousin, Raja Abdullah, and they ruled over Kota Jelasin. When they rejected the Siamese King's marriage proposal, they were attacked, defeated, and she surrendered to spare her subjects from further suffering. However, she assured her husband and subjects that she would return unharmed. When she met the Siamese King, he was inflicted with a terrible skin disease after touching her, curable only by Princess Sa'adong. She was allowed to leave, but discovered her husband had married another woman upon return, leading her to flee to Bukit Marak. Although she withdrew into the mythical world, she continued to assist her subjects, lending them crockery for feasts, which they failed to return, causing her belongings to turn into stones. Regretful, they returned her belongings to the cave in Bukit Marak, where some claim they can still be found.	1. Raja Abdullah's Tomb; Cave at *Bukit Marak*; *Istana Jahar*; *Istana Balai Besar.*

(continued)

Table 5.1 (continued)

State/ territory	The legends	Attractions
Sarawak	**1. The Legend of Princess Santubong** Princess Santubong and her sister, Princess Sejinjang, were sent by the King of Kayangan to bring peace between neighboring villages, Pasir Kuning and Pasir Putih. After succeeding, they ruled the villages, with Santubong overseeing Pasir Kuning's intricate fabric weaving, and Sejinjang managing Pasir Putih's thriving rice fields. Despite their beauty, they rejected numerous suitors until they met Putera Mahkota Serapi, who fell in love with both but could not decide who to marry. A quarrel between the sisters led to a physical exchange of blows. Sejinjang swung her thresher and hit Santunbong's cheek. As she fell on her back, Santubong threw her weaver and hit Sejinjang's head. As a result, they were cursed by the Kayangan King and transformed into mountains. Mount Santubong resembles a reclining woman, with a deep crevice at the peak symbolizing the thresher's impact.	1. Mount Santubong; Sarawak Cultural Village.
Sabah	**1. Monsopiad The Legendary Warrior** Monsopiad, born to the daughter of Kuai Village's headman, had a unique omen at his birth when Bugang eggs hatched on the roof of their house simultaneously, signifying his extraordinary abilities. As a young man, he was a natural fighter and vowed to protect his village from pirates who frequently raided. He single-handedly confronted and defeated them, ultimately beheading the ruthless pirate chief. Upon his return, he was celebrated as a warrior with a grand ceremony in his village.	1. Monsopiad Cultural Village
	2. The Kinabalu Dragon's Gem The legend tells of the Kinabalu dragon, which possessed a luminous jewel that caught the Chinese Emperor's attention. He sent his sons, Wong Wang Kong and Wong Song Ping, to obtain it. Wong Wang Kong gave up when he realized the challenge, but Wong Song Ping devised a clever plan. He created a lantern that imitated the jewel's glow and swapped it for the real jewel while the dragon was away. As they returned home, they were pursued by the dragon. Wong Song Ping had another ingenious idea: they heated cannonballs until they were red hot and launched them at the dragon. Longing for its lost treasure, the dragon swallowed one of the balls and sank below the water. Upon their return to China, Wong Wang Kong, consumed by jealousy, seized the jewel and presented it to his father. Disappointed, Wong Song Ping set sail, allowing the monsoon to guide him. Eventually, he reached the coast of Brunei, where he was warmly received.	2. Mount Kinabalu; Hot Spring.

	3. The Tale of The Stone Stump In a village near Sapulut River, there was a family of six boys and a beautiful girl. They often played near a large stone outcrop close to the river, home to a skunk or Tudoh with a foul stench they despised. Wanting to eliminate the Tudoh, they tried to cut down the stone but could not until the girl had a dream that they needed a shoulder blade, or Tulang Dayong, from a Landak Tunggal, a porcupine, to do it. They placed the girl on a nearby stone to keep her safe while they worked. As they struck the stone, their sister inexplicably grew taller with each blow. Eventually, the stone outcrop, with the Tudoh, gave way and fell into the Sapulut River, but the siblings were shocked to see their sister on top of the massive stone outcrop, crying for help. In their anger, they attempted to cut down the stone, but the peak broke off and flew all the way to Tenom Lama, around 50 kilometers away, known today as Batu Pinuto. Unfortunately, they lost their sister forever. The remaining piece of stone is now called Batu Punggol.	*3. Batu Ponggol;* Murut culture.
Labuan	**1. The Story of Corpse Island** Centuries ago, two young lovers faced opposition from their parents and decided to elope. Caught in the act, they were exiled from their village, and before leaving, they made a solemn vow that something dreadful would befall them if they ever returned. After enduring years of hardships, they prospered and remained deeply in love. However, as they grew older, they began to long for their family and decided to return to their village. Upon nearing the old village, the sky darkened, and a powerful windstorm generated huge waves that rocked their boat. The couple was cast into the sea, where a sea sprite, moved by their enduring love, transformed them into two small islands. These islands are now known as Pulau Mayat, or the Corpse Islands.	*1. Corpse Islands*; Labuan town in general

Source: Tourism Malaysia (2007)

and tales of Melawati Hill. However, legends and myths are also a type of folktales in the broad sense as explained in the definition provided earlier. Therefore, it is more advisable to adopt the broader term 'folktale or folklore' rather than legends or myths.

Besides promotional materials, the Ministry had also promoted these myths and legends through activities such as in the 2016 edition of the yearly festival named 'Citrawarna' (Colour of Malaysia). The theme 'Myths and Folklore' were adopted. "During this festival, states around Malaysia presented street performances based on themes associated with local myths and folklore stories (sic)" (Razak & Ibrahim, 2017: 10). Earlier on in 2001 the theme 'Malaysian Legends' were also adopted for the festival that year. The myths and legends presented were similar to those included in the brochures.

In 2016, the utilization of myths and legends in promoting attractions such as those in Langkawi has become less visible (Razak & Ibrahim, 2017: 1). In 2023 the website of Tourism Malaysia still carried the description that the islands of Langkawi are "shrouded with an intriguing heritage of myths and legends that feature ogres and gigantic birds, warriors and fairy princesses, battles and romance" (Tourism Malaysia, 2023). However, Langkawi is now portrayed as a natural paradise or classified by the UNESCO as "A Global Geopark" (Tourism Malaysia, 2023).

Langkawi Island can still be considered a role model in enhancing tourist experiences through developing tourist sites associated with myths or legends. These include the Legend Park, The Eagle Square, The Hot Spring Village, the Field of Burnt Rice, the Mahsuri's Tomb. Among those famous stories heavily promoted by the official tourism websites about Malaysia, Mahsuri's Tomb located at the cultural centre known as 'Kota Mahsuri' is one of the most significant attractions associated with myths or legends in Langkawi.

Presentations of these legendary sites to the visitors are however far from being satisfactory. One of the reasons is probably due to their state of poor maintenance. The Legend Park that cost Rm37 million to build in 1996 is a typical example (Razak & Ibrahim, 2017: 13). One may suggest to build a theme park to place all the intangible stories inside for tourists but this could prove to be very costly to build and to maintain. In the later discussion, we will explore other more cost effective ways to present the stories.

Besides, a new night trail of 1.2 km at Dream Forrest Langkawi was opened on May 1, 2023 which aims to provide a magical experience for visitors and tourists with features of characters and creatures from Malay legends like Merong Mahawangsa, Mat Teja, Mambang Sari and Gedembai among others" (Wee, 2023). Credits must be accorded to the Enfiniti Group president Puan Sri Tiara Jacquelina, the "Puteri Gunung Ledang" movie star who spent 4 years and Rm 20 million to make the night forest trails happen (Lee, 2023). Again, this is an expensive investment.

Although, legends are used extensively in promoting tourism in Malaysia, one is not difficult to notice weaknesses in the applications. First of all, the use is somewhat narrow, as only myths and legends are used extensively. I believe those folktales which have been included could have been made up by the Ministry of Tourism to fulfill the need that there is at least one story from each state. However, strangely, no story from Negeri Sembilan was included in the tourism booklet. Perhaps, the Ministry has found it difficult to get an appropriate legendary story from Negeri Sembilan that was formed from a federation of nine clans, which gave the state its present name (Portal Rasmi Kerajaan Negeri Sembilan, 2023).

As a side interest, I have often wondered why my humble, underdeveloped, little hometown in Pahang state, with a population of 3000 is called *Padang Tengku* (Prince Field). Indeed, such a name might have just come from an historical incident. I recently found out that, apparently, according to folklore, a local prince named Tengku Abdul Samah visited my hometown and was impressed by its open and vast landscape. He then decided to take a rest in the open field, thus giving it a place name *Padang Tengku* (Wikipedia, 2022).

These tales may allure tourists to them, especially those associated with places that are already attractive in other ways. One such example is the story about *Janda Baik* (kind widow or divorcee), a location in Bentong, Pahang which is popular for nature lovers and is not too far away from Kuala Lumpur. According to the story, a woman was abandoned by her husband who later married a new young girl and had a few children. One day they decided to go fishing by a lake near a jungle. The father went home early and the children got lost in the jungle, and found a kind woman who took them in and protected them from a man eater giant in the jungle. Upon learning about this event, the father was so regretful that he remarried the kind woman and lived happily ever after. The village they lived in was renamed *Janda Baik*. There are a few other stories about this kind woman named *Janda Baik* (Dris, 2016). Whichever version one learns about how this place was named, the legendary *Janda Baik* is attractive enough to draw in tourists who will further find the charming lake and its natural surroundings worthy of sightseeing.

There are many more such tales, such as how Teluk Sengat, Kuala Kangsar, Jementah, etc., got their names. The list can go very long. Puteh and Said (2008) have chosen 16 similar stories out of 366 stories collected. These stories can be used to allure tourists to these16 places. Puteh and Said (2008: 234) had collected other fascinating stories, one of which is about turtles that keep coming back to Rantau Abang, Terenganu to lay eggs on the beach while shedding tears. The two authors have other stories which explain how the giants designed the rocks in Pahang River near Batu Sawar village, and about how durian, the king of fruits for Malaysians got its pungent smell (Puteh & Said, 2008: 22). All these stories can be attractive to tourists.

4.1 Weaknesses of Malaysian Use of Legendary Stories

There is an over-emphasis on identical legendary stories which have been "told and retold over the years from one generation to the next … unfolded hundreds of years ago". No doubt old stories are indeed valuable. But according to Dundes (1965:1–3), stories which could be transformed into folklore may be created all the times by people from all walks of life such as taxi drivers, hoteliers, or simply people from the city street rather than from the rural setting. In Western countries, a particular genre is about urban ghost stories which have often attracted some attention in recent years. Such legendary stories have potential to draw in curious visitors to take up tripping or dark tourism to haunted places where hotels have been built and become popular (Holloway, 2010; Ironside, 2018). Using ghost stories may be a little sensitive to promote tourism in a religious country like Malaysia, though it may serve as a possible alternative to lure tourists in the cities.

Another possibility is by means of organizing cultural shows. A good example is the 'Folklore by the Sea', a live theatrical set in the heart of the famous Chew Jetty where visitors are told the oral history of Chew Jetty Penang (Tripadvisor, 2024). As a multi-racial society, it is a pity that the Ministry of Tourism rarely includes non-Malay folklore stories, probably because non-Malay legends might not be old enough to be cited or the legend collectors are themselves unfamiliar with non-Malay legendary stories. While I am less familiar with Indian folklore in Malaysia, the Chinese folklore researchers in Malaysia have certainly collected many local folklore stories (Lee, 2019). In this sense, we see the need that the Ministry of Tourism should put in greater effort to make non-Malay stories available in Malay and English language. It is also good to have these non-Malay promotional materials exposed in other major international languages such as Arabic, Chinese, Japanese, Korean, etc.

Having said that, after selecting the tales, Tourism Malaysia should promote them online, in pamphlets or any other means, and provide facilities for tourists to link their experience with physical objects, thus creating a visualization effect. For those locations that have a natural site, building or object associated with the tales, they will be attractive enough to appeal to the tourists for a visit. For tales having no such physical association, facilities such as legend park, LED lighting trail or museum may be built to present them. Management of such a park may involve big investments and is expensive to maintain, careful planning is duly needed.

A much cheaper facility is mural. Murals have gained much popularity in Malaysia in recent years, starting from those found in George Town, a UNESCO world heritage. Many tourist spots have started with painting murals. Its popularity is due to the current social media culture. Many tourists would like to take picture to be posted in social media to record their visit. However, most murals found in Malaysia are randomly painted, lacking in any theme. In fact, murals can be painted according to local tales which is meaningful. There are already such mural cases in Langkawi (Razak & Ibrahim, 2017: 14) and in other sites in Asia (Wannakit, 2018: 75). Mural artists should be encouraged to do so, and what they do will help to educate tourists about the local culture.

To help propagate legendary stories to locals and visitors, modern 'active bearers of traditions'[1] such as tour guides, hoteliers, taxi/Grab drivers should be made familiar with local tales, so that they too can tell them to the tourists while visiting

[1] A term some folklorists use to describe people who transmit folklore.

the sites. I enjoyed the stories my professor friend told me while walking in the streets near his house in Tai Po, Hong Kong so much. Having such knowledge has allowed me to understand Hong Kong in much more depth.

5 Conclusion

Tourism Malaysia and the Ministry of Tourism Malaysia have made good use of Malaysian myths and legends to promote tourism in Malaysia since the 1950s, especially to promote Langkawi Island. However, their use in this contact is not without weaknesses.

First of all, the stories selected for tourism promotion are mostly old Malay or indigenous tales, with a few exceptions related to colonial history. There is an overemphasis on identical legendary stories, and the use of non-Malay folklore is limited. Some of the legends selected are folktales rather than legends or myths, and the term 'folktale'is suggested for a broader description. In fact more folktales are encouraged to be included in promoting tourist sites. These stories can be promoted in various ways, including through festivals like "Citrawarna."

Secondly, facilities like legend parks, LED lighting trails, museums at some legendary sites have not been well-maintained, possibly affecting their presentation to visitors. Understandably, these are expensive investments. More cost friendly alternative approaches to promote folklore such as cultural shows, story telling, multimedia shows, murals etc., should be explored.

Thirdly, local guides and other tourism professionals should be educated about these tales to share with visitors.

The potential using folklore to attract tourists to Malaysia is very good. Therefore, we should be more encouraged to explore further, wider and deeper into the efforts of making full use of local folklore to promote and enhance tourists' experience for tourism in Malaysia. We can also learn mutually with our counterparts from other countries or regions in a joint effort to consolidate our folklore knowledge and experience which will definitely contribute towards our tourism industry.

References

Dris, M. (2016). Asalnya Kampung Tiga Haji, ditukar jadi Janda Baik. *Berita Harian*, Feb 7, 2016. https://www.bharian.com.my/bhplus-old/2016/02/121964/asalnya-kampung-tiga-haji-ditukar-jadi-janda-baik

Dundes, A. (1965). What is folklore. In A. Dundes (Ed.), *The study of folklore*. Prentice-Hall.

Dundes, A. (1975). *Analytic essays in folklore*. Mouton Publisher.

Geraldine, A. (2023). When village folklore meets tourism. *The Star*, June 5, 2023.

Holloway, J. (2010). Legend-tripping in spooky spaces: Ghost tourism and infrastructures of enchantment. *Environment and Planning D; Society and Space, 28*, 618–637.

Ironside, R. (2018). Chapter 5: The allure of dark tourism: Legend tripping and ghost seeking in dark places. In D. Waskul & M. Eaton (Eds.), *The supernatural in society, culture and history* (pp. 95–115). Temple University Press (online).

Lee, E. K. (2019). *Zhuanchi huaren de eyu: Folk tales of old Taiping*. Self-publishing.

Lee, M. (2023). Dream forest Langkawi: An Rm20mil night forest trail attraction opening in may. *RojakDaily*. Feb 28, 2023. https://gempak.com/rojakdaily/lifestyle/dream-forest-langkawi-rm20mil-night-forest-trail-attraction-opening-may-79711

Penjore, D. (2005). *Folktales and education: Role of Bhutanese folktales in value transmission*. Paper presented at the second international conference on Gross National Happiness, Local Pathway to Global Wellbeing, St. Francis Xavier University, Antigonish, Nova Scotia, Canada June 20 to June 24, 2005.

Portal Rasmi Kerajaan Negeri Sembilan. (2023). *The early history of Negeri Sembilan*. https://www.ns.gov.my/en/kerajaan/info-negeri/sejarah-penubuhan

Puteh, O., & Said, A. (2008). *A collection of Malaysian folk tales*. Utusan Publication.

Razak, N. A., & Ibrahim, J. A. (2017, December). From names of places to Mahsuri's curse: Exploring the roles of myths and legends in tourism. *International Journal of Business, Economics and Law, 14*(2), 10–17.

Razak, N. A., & Romle, A. R. (2016). Representation of native myths and legends and cultural values in the Malaysian tourism promotional brochures. *Research Journal of Applied Sciences, 11*(11), 1379–1383.

Richards, G. (2003). What is cultural tourism? In A. van Maaren (Ed.), *Erfgoed voor Toerisme*. National Contact Monumenten.

Thomson, S. (1946). *The folktale*. Rinehart and Winston.

Tourism Malaysia. (2007). *Treasured Malaysian legends*. Ministry of Tourism Malaysia.

Tourism Malaysia. (2023). *Langkawi- Welcome*. https://ebrochures.malaysia.travel/langkawi-the-jewel-of-kedah/langkawi-welcome/

Tripadvisor. (2024). *Folklore by the sea*. www.tripadvisor.com/Attraction_Review-g298303-d15684943-Reviews-Folklore_By_The_Sea-George_Town_Penang_Island_Penang.html

Wannakit, N. (2018). Application of folktales to cultural tourism management: A case study of the Central Isan Provinces of Thailand. *Manusya: Journal of Humanities Regular, 21*(1), 66–84.

Wee, J. (2023). Malay folklore to come alive in unique immersive experience at Dream Forest Langkawi in May. *Malay Mail*. Fed 23, 2023.

Wikipedia. (2013). *Ong Sun Ping*. https://en.wikipedia.org/wiki/Ong_Sum_Ping

Wikipedia. (2022). *Padang Tengku*. https://ms.wikipedia.org/wiki/Padang_Tengku

Wong, J. (2011). Chinese hero in Sabah Folklore?: Part 2 So who is Wong Song Ping. *Jsunamwong's Blog*. May 5, 2011. https://borneonfarmer.wordpress.com/2011/05/05/chinese-hero-in-sabah-folklore-part-2-so-who-is-wong-song-ping/

Hoon-Peow See holds a PhD in Sociology and Anthropology from the International Islamic University Malaysia, a Master of Philosophy in Sociology from the University of Malaya, and an LLB (Hons) from the University of Glamorgan (now the University of South Wales). He has held various leadership positions in the education industry with exposure in both academic and business operations, which include being CEO cum Vice Chancellor of a university college. He is currently the Director at Sunway TES College Johor Bahru, Malaysia. He has a wide academic interest in humanities and social sciences. Although he is a full-time manager, he is active in research and publishing. He has published many books and papers locally and internationally with reputable publishers and journals. He has also presented papers at local and international conferences.

Chapter 6
Examining the Trade-Offs of Visiting Fee Rises to National Parks, Sabah, Malaysia

Tai-Chee Wong

1 Introduction

Kinabalu Park is Malaysia's national park in Sabah which was first gazetted in 1964 as a natural protection site in appreciation of its affluent and diverse plant and animal life. In December 2000, the Park was declared by UNESCO as Malaysia's first World Heritage Site for its "outstanding universal values" with very important biological assemblage of tropical plants and animal species in the planet. The Park stands across two mountains, Mount Kinabalu (4095 m-the highest in Southeast Asia) and Mount Tambuyukon (2579 m), and it covers an area of 754 km^2 (MountKinabalu.com, 2023; Carrier & MacLeod, 2005).

Before 2023, Kinabalu Park was open to visitors by charging a small admission fee as a token to pay off administrative cost and help in conservation efforts. Effective from January 2023, however, the entrance fee to the Kinabalu Park had been raised more than three-fold from 3 Ringgit to 10 Ringgit[1] for local adult visitors, and Rm 50 to Rm 100 for climbers (MountKinabalu.com, 2022). Fundamentally, the rationale underlying the rise serves as a means to discourage fast rising numbers of visitors who are liable to bring damages and disturb the regenerative capacity of plant and animal species, seen as harmful to the environmental sustainability of the Park. Use of higher fees and travel restrictions as a management strategy and policy to safeguard protected areas is not new. Similar restrictive measures were earlier adopted by Bhutan to avoid tourist overcrowding to its mountain resorts, and in Australia, by the state government of Victoria to restrict the number of visitors

[1] As of July 2023, one US dollar was equivalent to 4.6 Malaysian Ringgit.

T.-C. Wong (✉)
New Era University College, Kajang, Selangor, Malaysia

T.-C. Wong et al. (eds.), *Cultural Tourism in the Asia Pacific*,
https://doi.org/10.1007/978-3-031-63459-8_6

attending the daily "penguin parade" on the Philips Island in order not to disturb the colony's routine activity (Llausas et al., 2019).

Visits to national parks require, in essence, precautions in ecological preservation and protection. In this sense, it involves necessarily careful eco-touristic measures that have to seek a check and balance in restricting visitor impact yet providing rights to the people in their admiration of natural beauty and its openness to fresh air and sunshine. Arguably, there is here an aspect of sustainable tourism development where a triangular relationship is at stake. This stake is the search of a balance between environmental justice, governance and carrying capacity in relation to the park in question (see Clayton, 2002; Borrini-Feyerabend & Hill, 2013). This triangular relationship will be discussed further in a section below.

As shown in Fig. 6.1, setting an optimal number of visitors may be sought as an ideal solution based on the assumption that as a result of fee hikes there will be a reduction of visitors and less revenues and less resources for nature protection. However, less visitors means also less detrimental effect and ecological footprint which is good for nature conservation. Striking a balance by setting a number from a mathematical model is however seen to be unrealistic and is highly controversial (see Llausas et al., 2019), we argue therefore on the basis of carrying capacity, environmental justice and governance which all have an impact on the achievement for an ideal optimum of tourist visits. This analytical methodology is merely a conceptual conflict resolution approach and it does not lead to any idealistic numerical projection of visitors which is not the objective of this paper. Based on this rationale, the paper examines the check and balance with regards to the triangular relationship between environmental justice, governance and carrying capacity through which a conceptual setting is made for an ideal and optimal target of visitors (see Fig. 6.1).

2 Tourism Sustainability: The Triangular Relationship of Carrying Capacity, Environmental Justice and Governance

Earmarking ecologically fragile parks as national parks yet allowing limited visitors is an integral constituent of ecotourism. This approach is characteristically different from conventional mass tourism given its nature of responsibility beneficial to environmental sustainability with an objective to minimize the degrading impact to the environment. Yet, ecotourism, if abused or being used like a "bursting bubble", may be turned into a form of environmental exploitation, not much different from mass tourism (Carrier & Macleod, 2005). It is thus intriguing to address the relationship between the concept of carrying capacity, environmental justice and governance in this context of national park access issue.

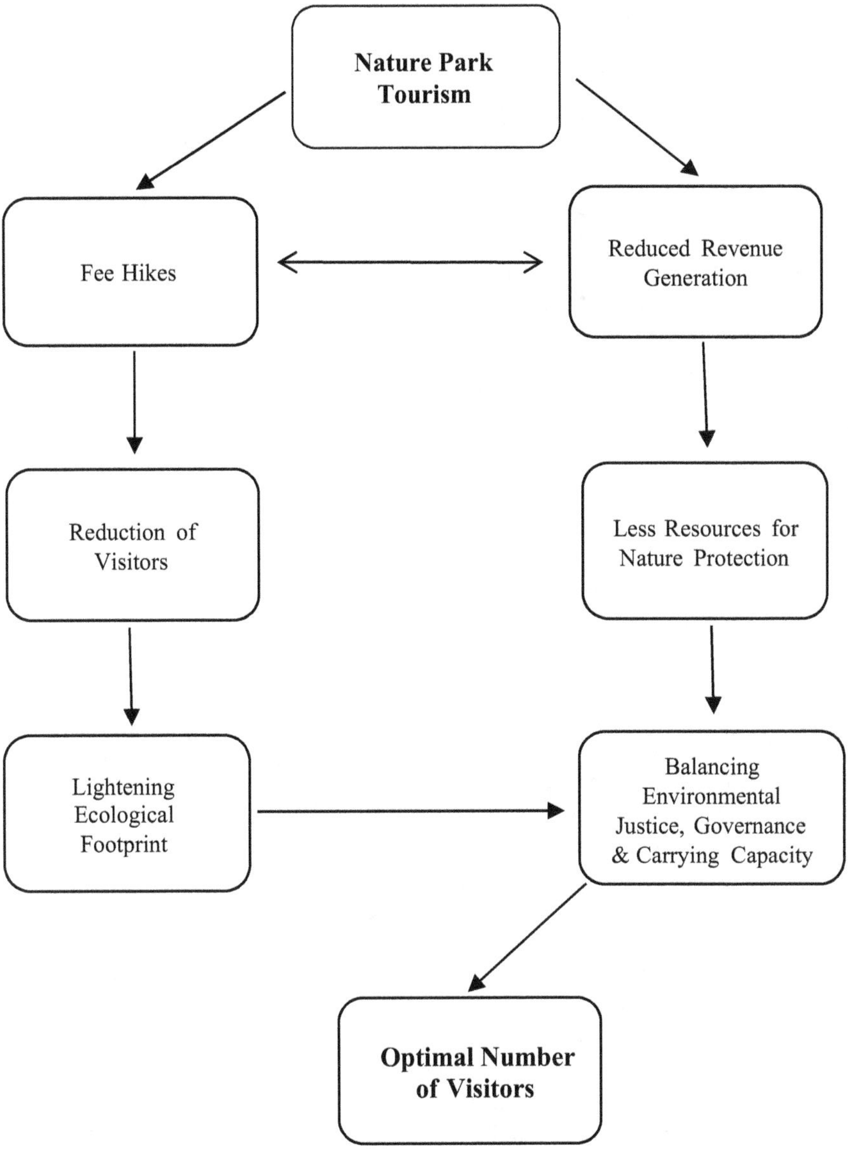

Fig. 6.1 Ideal conflict resolution between fee hikes and nature conservation. (Source: Conceptualized by author)

2.1 Carrying Capacity

Carrying capacity is usually seen as an inherent upper limit of ecosystems that tolerates exploitative human activities to achieve sustainable rates of natural resource use. It can also be defined as the maximum number of a given species that a given environment can support without experiencing degradation (Seidl & Tisdell, 1999). Evolving from the study of biological sciences, the notion of carrying capacity is commonly used in the livestock and forestry industries where maximum production is sought but future yields should be sustainably maintained. More recently over the past few decades, the term carrying capacity has been applied upon the relationship between human activities and natural environment. In an ecological sense, the carrying capacity concept implies the number of visitors an area is capable of supporting in a sustainable manner without degrading natural resources within it and its immediate surrounding areas. In the tourism aspect, the social carrying capacity of an area refers to the level it possesses in recreational value which has the capacity to fulfill the expectations of visitors without threatening its environmental sustainability in the long term (Sievänen, 2004; Clayton, 2002).

Putting the ecological and social carrying capacity factors together, there is a critical managerial implication that attitudes and preferences of visitors as rightful users have to be considered and made compatible with an effective managerial planning. The ultimate objective of these two factors combined is to minimize ecological impact yet fulfill tourist satisfaction to an acceptable extent. In the nature-based tourism comprising nature park visits, the concept of limits of acceptable change (LAC) has developed a principle which encompasses a compromise between ecological, economic, and social dimensions. The LAC assessment process may be applied as a tool to evaluate the impacts of recreation on natural areas whereby visitor conflicts and other related problems could be managed (Sievänen, 2004).

The carrying capacity of an area may be mathematically computed based on certain assumptions though its reliability and accuracy have remained doubtful. In this aspect, de Sousa and his research team worked on the carrying capacity estimation on three Brazilian estuarine beaches, known as Colares, Maruda and Munuubira beaches. In their study, their carrying capacity estimation was computed using three elements for consideration, and these include the physical carrying capacity (maximum number of users that can be accommodated within a certain area within a specific time), the real carrying capacity (maximum tolerable number of people within a certain area taking into account ecological, social and management features of the site) and the effective carrying capacity (maximum number of visitors the site can support after considering the real carrying capacity and the management capacity inclusive of infrastructure, services and equipment available on site). Based on this methodology, the research team found the maximum daily carrying capacity of these three beaches at 674, 812 and 97 visitors respectively (de Sousa et al., 2014).

The notion of carrying capacity on beaches may well be applied to nature parks, as well as to a broader context such as towns, cities where the provision of water supply, electricity, sewage treatment systems as well as community services, each of which may serve as a limiting factor to an ideal urban growth. In assessing the carrying capacity of a nature park open up to visitors, the concept has required a basis for optimizing economic gain and environmental conservation at a level where the sustainability of both tourism and ecological preservation is not compromised (Clayton, 2002). In other words, the aggregate demand of the tourist population may be limited by the carrying capacity yet economic gain must not be adversely affected.

2.2 Environmental Justice

Access rights involve the imposition of control measures to public access with an aim to protect the ecological balance of protected areas for a sustainable future. Forbidding public access rights therefore infringes environmental justice. In some parts of the world, inhabitants were sometimes forced to vacate an identified protected area which was converted to a national park, and such an act may be classified by human rights groups as an "environmental racism" (Dahlberg et al., 2010; Brockington & Igoe, 2006).

Once converted to a national park, control of access is almost automatically superimposed in the name of site protection. In the more developed capitalist world, many governments and conservation activist groupings including the non-governmental organizations, access policies have often been revised with an objective to reconcile conservation measures and environmental justice. This issue may be rendered complex by incorporating arguments inclusive of the terminology usage such as "power, scale, equity and human rights" in the context of political ecology and democratic accountability (Kothari, 2008; Dahlberg et al., 2010: 210).

Arguably, no matter how complex the social values in the intellectual debates on power relations, equity of access and democratic accountability, the ultimate goal of the issue is after all to achieve sustainable management of nature parks as part of land resources. In this sense, environmental governance with local engagement is a prerequisite that we should look up to.

Denying customary rights to the local community in access to national parks, once gazetted for conservation, is controversial. This is particularly so if the surrounding indigenous inhabitants are dependent on harvesting certain forest products for a livelihood. The conserved sites may also comprise of old ancestral settlements found with graves and shrines and other geographic features worshiped as respectable and significantly mythological spirits. Occasionally, certain locations within the conserved zone might have been used periodically for the performance of rituals and ceremonies (O'Connor et al., 2013). Such sensitive issues are required to be resolved harmoniously.

As such, the question arises as to whether the evasion of indigenous inhabitants from the gazetted park constitutes an infringement of their rights, including that of depriving them of their right to derive a livelihood.

2.3 Governance

Most kinds of human activities have created adverse impacts on nature and this has required governance via policy measures to rectify the adverse consequences that tend to harm the sustainable future of human society. By governance, we have in mind the need to pursue an optimal and appropriate level in the management of protected areas, one of which are national parks whereby we seek in essence a balance between the requirements of human-led economic development and those of conserving biological diversity (Borrini-Feyerabend & Hill, 2013: 171; Cardenas et. al., 2015).

To these intents and purposes, the United Nations enacted the International Union for Conservation of Nature IUCN in 1948, by providing it with a consultative status at the organization and an objective designated at nature protection and conservation. It has since become the world's largest environmental network comprising over 1500 member organizations and a large number of experts serving six commissions dedicated to species survival, environmental law, protected areas, social and economic policy, ecosystem management, and education and communication (IUCN, 2021). As we can see, all six commissions are in one way or another associated with nature park conservation efforts. IUCN's global mission includes organizing forums where governments, non-governmental organizations (NGOs), scientists, business and community representatives including indigenous groups are involved to work together to forge solutions or compromises to environmental challenges. By its global responsibility in ensuring transboundary nature conservation, IUCN has provided a useful voice in terms of international governance in environmental conservation.

Having been criticized for placing the interests of nature and closer relations with the business sector over those of indigenous peoples. IUCN has in recent years shifted its conservation project portfolio by combining both most-up-to-date scientific approaches and the traditional knowledge of local communities in their efforts to "reverse habitat loss, restore ecosystems and improve people's well-being" (IUCN, 2021). By so doing, the IUCN has viewed that getting the involvement of the indigenous people is not only part of a democratic process that helps drive its global conservation agenda, it is also an initiative to "recognize and support indigenous peoples' contributions to the conservation of biodiversity and climate solutions through their indigenous knowledge systems". In this aspect, this new initiative placing great importance on indigenous knowledge has envisaged four pillars of action as follows (IUCN, 2023):

(a) Capacity building of the indigenous people;
(b) Scaling up work with indigenous peoples at the territorial level;
(c) Empowering indigenous peoples-led design in the development of direct funding approaches; and
(d) Recognizing and maintaining indigenous people's leadership in international spaces and platforms.

In this regard, we examine below the governance issues addressed by the IUCN in a number of aspects. Firstly, nature parks to be conserved must be defined with a specific geographical space with a borderline so that enforcement of rules and regulations could be made effective. Then, what criteria should be used to define such borderline for management purposes? Secondly, who should draft up an appropriate natural resource rules for management purposes and what is the manager's scope of decision in interpreting such rules and regulations? Thirdly, with regards to the assessment of the quality of protection measures, what are the indicators to be used and how should the monitoring and evaluation procedures be institutionalized and enforced? As such, there are many more such challenging issues that need to be addressed to fill up the gap of governance requirements (see Borrini-Feyerabend & Hill, 2013).

Undeniably, there is a diverse means of nature parks governance via different institutions in open societies. These include, among others, governance by various levels of government, governance by various rights-holders and stakeholders, called shared holders, governance by private individuals and organizations, and governance by indigenous people and/or local communities (ibid: 180; Törn et. al., 2008). As a large proportion of national parks worldwide are closely associated with the survival interests of indigenous groups, how then traditional knowledge and indigenous people's involvement could contribute towards the sustainability of nature areas including biodiversity conservation deserves an in-depth inquiry. For this purpose, we wish to refer to a case study in good detail below where we explore how indigenous ways would work in order to avoid humans' transforming nature in reversible paths to the extent of no return.

2.3.1 Governance by Indigenous Peoples/Local Communities

To a great extent, an indigenous people and a local community are interchangeable when their existence in a specific area is long-standing and has established an inseparable and interdependent relationship with the nature area from which they derive partly or wholly subsistence materials for their living needs. In this sense, the indigenous community would have settled there, managing land, water and natural resources collectively in a sustainable manner and practicing often their customary collective rights over these resources. This is required as they have to develop their livelihood strategies in simple ways in response to opportunities provided by nature and resolve challenges by fitting local ecological conditions. They practically do not try to alter natural conditions in a substantial way, and thus are contributors to the "conservation of ecosystems, species and ecosystem-related values" (Borrini-Feyerabend & Hill, 2013: 183).

Characteristically, opening up a national park for free public access has both positive and negative effects towards environmental preservation and the local community. A study by Jaafar, Ismail and Rasoolimanesh (2015) on the tourism impact of Kinabalu National Park has revealed that most of the local residents surveyed indicated their acceptance of tourism activities around their habitat. Local indigenous groups admitted that tourism had acted as a stimulus which had improved their general standard of living and quality of life. Negative effects they highlighted were "congestion and the exposure of anti-social behavior to the local community". By anti-social behaviour, it means clearly foreign visitors, notably those from liberal Western nations may show acts that are taboos to local traditional practices. The study has also discovered that the surveyed population felt tourism had weakened the community's social bonds and family structure. Overall, however, their perception was that positive effects outweighed negative effects (ibid: 5, 14).

In recent years, there has been an emphasis by Western intellectuals placed on nature-culture relationship linked to the customary "tenurial rights" of the indigenous communities in managing their ancestrally given homelands (O'Connor et al., 2013). However, the pressing issue is how good the quality of environmental governance of indigenous people or local community in managing their own lands, given their generally low levels of formal education with scientific training despite their centuries of traditional experiences. Mainly because of this weakness in formal scientific training, ethnic-based local organizations are often not formally recognized by the state in governing conserved areas. To help enhance the capacity of local communities in achieving the ecologically sustainable parameters set by scientific institutions, and thus their governance, a few approaches may be sought to strengthen such governance, suggested as follows:

(a) Sustainability knowledge and consultations: Local conservation community leaders should be trained to get accustomed to understand an organized system of governance. This includes the responsibility and accountability when holding authority to make decisions in accordance with principles of acceptable governance. Good governance implies quality decisions to be made with good knowledge in environmental sustainability issues and consultative collaboration with the public authorities and local representatives;
(b) Wise governance: Wise governance is more than "good governance" at the implementation level. A functionally workable team should be gathered to work harmoniously and effectively, and the team's motivation should be based on a strong will to serve the common good which encourages care, generosity, patience, humbleness and discourages selfishness and greed; and
(c) Empowered governance: Here it involves a decision-making system that enables local communal organizations to organize their responses to changing environmental conditions promptly on the spot based essentially on their localized knowledge. This capacity building helps in confidence building and is conducive to enhance leadership quality. It is also a way of recognition over the assumption of responsibility of local leaders (Borrini-Feyerabend & Hill, 2013: 196–198).

3 Case of Sabah's Kinabalu Park and Conservation Measures

An overview of the above discussion over the triangular relationship of carrying capacity, environmental justice and governance should have given us a basis of conflict resolution. Historically, conservation efforts have taken substantial amount of time and conceptualization in the contemplation of what is best to protect national parks. Originally, national parks were not identified predominantly for the sake of environmental conservation and site protection. For example, in 1872 the Yellowstone Park bill was passed by the American Congress more to be:

> dedicated and set apart as a public park or pleasuring ground for the benefit and enjoyment of the people ... [and] to provide for the preservation, from injury or spoliation, of all timber, mineral deposits, natural curiosities, or wonders within said park, and their retention in their natural condition (Mackintosh et al., 2018: 6–7).

As polluting sources increasingly accumulated over the industrialization age in the last 150 years, protective measures were taken as a remedy to safeguard human living conditions, diffusing initially from the Western industrial world to the less developed regions. Given the inseparable boundaries of pollution impact over space, the scope of environmental protection has spread from urban-industrial sites to include rural and natural areas. Under the leadership and sponsorship of the United Nations, the notion of "one mother earth" has eventually convinced national governments to accept the need to share the joint global responsibility to protect our planet.

Within this context, national parks have been increasingly identified on the basis of their vulnerability to human impact and their related activities. The agenda of nature preservation had reached Malaysia and influenced its conservation policy in protecting its rich tropical flora and fauna rare species under threat. The Endau Rompin National Park serves as a good example where 52% of Malaysia's 286 rare mammalian species are found. While such species are attractive to wildlife tourists, the species themselves should be a target for sustainable protective measures. In a study conducted by the Forest Research Institute Malaysia in 2016, the visitor respondents to the Endau Rompin National Park, located in Pahang state in West Malaysia, expressed most noticeably their strong interest in observing endangered species and rare mammals. They said they most enjoyed their jungle trekking and camping experience (Aihara et al., 2016: 177–178). Whilst we appreciate their positive response, we feel that it has also provided us with an inductive implication that control of visitor numbers is highly useful in terms of rare species protection.

Compared to the Endau Rompin National Park, Kinabalu Park in East Malaysia stands out even more significantly with its four climatic zones, housing one of the world's richest collections of biodiversity with over 4500 species of flora and fauna where 326 bird species and over 100 mammal species are found. At this time period, nine walking nature trails have been planned in the Kinabalu Park, one of which leads to the summit Mount Kinabalu. Climbing this highest mountain in Malaysia must be pre-booked and a daily limit of visitors is imposed. Wildlife watching including bird watching is a key interest of many climbers and visitors.

3.1 Unique Properties of Kinabalu Park

Kinabalu Park has a unique naturally functioning ecosystem in which ideal tropical conditions are found in support of its vast diversity of biota and high endemism, ranging from lowland rainforest to alpine scrub towards Mount Kinabalu's peak around 4000 m at mean sea level. The Park is species-rich containing an estimate of 5000–6000 vascular plant species, notably those classified under the families of all flowering plants. These include 1000 orchid species, 78 species of ficus, and 60 species of ferns, among others. There are a total of 326 bird species, as well as 90 species of lowland mammal and 22 species of mammal in the upper montane zone.

3.2 Protection and Management Measures

Legislation and institutional regulations of Kinabalu Park were adopted under the Parks Enactment 1984 and Amendment of 2007 with the objective to protect the park by specifying its functions, procedures and control measures. By management and ownership, Kinabalu Park is placed under the jurisdiction of the State Ministry of Tourism Development, Environment, Science and Technology and the State Ministry is subject to observe legislations passed down from the state and federal governments. Nevertheless, the Federal government's National Parks Act does not apply to Sabah which is solely responsible for managing and enforcing legislative measures specific to Kinabalu Park (UNESCO, 2023).

By Southeast Asian standard, Kinabalu Park's recruitment of competent staffing and budget levels are sufficient to maintain a reasonably high level of management and conservation. Essentially, the key issue is the rising pressure from commercial tourism, and the encroachment of adjacent land areas for other uses. Much of the lowland forested areas near the Park have already been transformed to other agricultural uses and developments. Hitherto, the Sabah government has closed mining and logging activities bordering the Park, and enforcement has been effectively controlled.

Commercial tourism constitutes the crux of the conservation measures and the rising ticket charges are part and parcel of the discussion in this paper as to where to strike a good balance. As such, the setting of the upper limit of tourism has remained a thorny issue. As to where to set the visitor margins and what facilities to provide, this has required consequently an extensive planning and public consultation to ensure minimal impacts within the park as and when the number of visitors increases. Other measures that have been undertaken to preserve and protect the existing flora and fauna are as listed below (UNESCO, 2023):

(a) It is an offence to remove or pluck any plants and to shoot, trap or collect animals;
(b) Rubbish must be disposed of appropriately and voluntary assistance to pick up rubbish and dispose of is welcome;
(c) Bringing pets is not allowed as they may carry diseases and harm the habitats inside the park;
(d) Do not disturb, making inappropriate noise, and movement when confronted with animals; and
(e) Do not bring packaged products which may result in intentional or unintentional littering.

4 Conflict Resolutions

In general, conflict resolution studies are largely value-based more focused towards political science, and it serves largely as a "mechanism of peace building, peace building, peacemaking and peace keeping [which] includes only peaceful methods and techniques for the maintenance of peace and security and to protect the succeeding generations from the scourge of war" (Wani, 2011:104).

In the environmental preservation aspect, conflicts arise from disagreements among stakeholders over disputes involving protection of environmental quality or natural resource management issues. Conflicts could be commonly grouped into six categories, namely "land use, natural resource management and use of public lands, water resources, energy, air quality and toxics" (Nicholson, 2009: 2). The nature of conflict may be further categorized into site-specific cases for which resolution measures may be subjected to national policy and the relationship and priorities set by the central authorities and the local governments.

In normal practice, resolution of environmental conflicts involves the seeking of a balance between popular demand towards the use of the preserved object and the need to assure the minimal standard to protect the natural state of the environment. The means employable to resolve the environmental or natural resource issue may be through mediation or negotiations between stakeholders to reach a compromise. Conflict resolution is unlikely to work if one party sees no good reason to compromise. In a democratic society, it would be most ideal to rely on a consensus-based approach to resolve disputes. Often, the process allows governmental agencies and the local stakeholders to mediate and negotiate over the affected interests in a democratic decision-making deal (Törn et. al., 2008; Cohn, 2002). NGOs may be involved and their role if well played objectively could bring about a less biased resolution. A site-specific case of Mantanani Island is cited for a deeper discussion of an environmental conflict resolution.

4.1 Sabah's Mantanani Island and Conservation Tourism

The type of conservation tourism put in practice on Mantanani Island is obviously classifiable in the context of ecotourism. Ecotourism to be genuine must be very different from conventional mass tourism which is liable to bring along unfavourable consequences harmful to the local natural environment. Ever since the United Nations spectacularly declared 2002 as the International Year of Ecotourism, ecotourism sites which are often identified as "indigenous" or "exotic", carry with them the notion of responsibility and respects expected from ecotourists to support their conservation efforts. Most frequently, ecotourists are likely to pay user fees towards the conservation projects, help empower local communities and to be educated in environmental principles while enjoying their visit of natural areas. Sadly, what is happening has drawn critical attention that "ecotourist bubble" is being witnessed when business is expanded to ecotourist spots which attract numerous international tourists (Carrier & MacLeod, 2005: 315–317).

Mantanani Island is one of the 394 islands off Sabah's west coast with rich marine life and biodiversity. Visitors are for the most part attracted by the island's scenic beaches, clear water and rich marine life ideal for diving and snorkelling. Before the outbreak of the Covid-19 epidemic in 2019, Sabah received 3.9 million visitors with a turnover of Rm 8.3 billion in revenue, making tourism the second largest source of state income. In light of its good potential in ecotourism, a sustainable tourism development plan has been put in place to protect its natural heritage and indigenous inhabitants with their unique culture, identity and history (Sulehan & Gabir, 2021). The dominant indigenous group living on the island are the Bajau Ubian who are known as skilled sea dwellers. Their main source of livelihood is traditionally fishing. With the development of ecotourism on the island, they have joined tourism business as resort employees, operators of local cafes, restaurants, provisions and souvenir shops as well as acting as local tourist guides.

Hence, the sustainability issues involved in the development of ecotourism on the island are two-fold in two respective environmental and economic senses. The first is ecological preservation of the natural state of the island in the face of the growth of community-based and external-sourced private tourism businesses.The second is how we could ensure the indigenous islanders could benefit from the ecotourism development while their traditional livelihood is being disorientated. More importantly, with rising population and influx of tourists, can ecotourism thus developed be made to uphold and sustain local values and traditions?

A conservation plan drafted in 2019 by the Sabah state body, Sabah Park, has addressed the need for a conflict resolution which focuses on the practice of a balanced development strategy. This strategy aims to ensure there would be no overexploitation by the tourism industry which runs the risk of destroying the marine diversity as a result of its pollutants such as garbage and sewerage disposals. The safeguarding measures involve the use of income generated from tourist fees and other sponsored revenues.

Over-fishing is another concern. A survey conducted by Sulehan and Gabir (2021) has indicated that reactions from local fishermen were mixed when asked if they were agreeable to the fact the island would be gazetted for environmental conservation but fishing activity would be made highly restrictive. The survey shows four of ten interviewees objected or unsure of the Sabah authority's motive in prohibiting them in fishing. In the face of the impact of ecotourism which has a strong tendency to change the source of livelihood of the islanders, the resolution will inevitably have to rely on getting alternative jobs. In this case, ecotourism will play a key role in it. Yet, economic planners will have to tackle issues of the islanders who lack basic skills and capital to start any business pertaining to ecotourism. Without proper skills training, local youths are likely to fall into cyclical unemployment and consequently poverty (ibid: 44–47). If poverty persists among most of the islanders who have chosen to stay behind, the risk of not getting cooperation from local inhabitants in the enforcement of nature conservation measures is high!

We now come back to the issue of conflict resolution between fee hikes and nature conservation of the Kinabalu Park. Having given a conceptual consideration to this issue in terms of carrying capacity, access rights and governance, it is felt that fee hikes are indisputably important to raise revenue to help in the conservation efforts. Logically, fee hikes tend to reduce the number of visitors and would therefore tend to lower the detrimental effects to the quality of Kinabalu Park's natural environment. However, this logic may be seen as simplistic if management governance overrides fee hikes with quality services and conservation expertise.

The dilemma however is not where to set the cut-off point or finding the "magic number" as the "scientifically optimal" number of visitors. Often, the validity of assumptions used in such scientific research has been challenged, given the highly complex dynamics in the natural ecosystems. In actual practice, moreover, conventional scientific ecological knowledge may only play a minor role in the decision-making process. Other factors involving the interplay of different stakeholders such as business interest, uneven distribution of power among stakeholders, weaknesses in enforcement and management expertise and efficiency could be more influential in rendering distorted decisions. In the final analysis, the conflict resolution must be complemented with a social and biopolitical context in setting the goals of planning and implementation (Llausas et al., 2019).

5 Concluding Remarks

Once a nature area is open up to human traffic and tourism-related activities, it is virtually impossible to rule out totally adverse ecological effects on nature's original state and the consequence that brings about an extent of dispossession and alienation to the existing flora and fauna and local inhabitants. Pollution and contamination of local substances are bound to take place and it is only a matter of how extensive and intensive the impact would be (Carrier & Macleod, 2005). To minimize the impact, collaborative efforts from the combination of stakeholders are critical. These include appropriate policy design of environmental conservation, thoughtful tourists who are highly conscious of their duty to protect the park, and to achieve a striking balance in setting a limited and optimal number of visitors. This combination of factors has the capacity to turn environmental exploitation into environmental protection.

Nevertheless, there remains a conflicting and controversial issue as to what could be accepted as optimal number of visitors. Using the carrying capacity construct with technical assumptions to set optimal numbers justified by research methods tends to invite critique of the concept and its suitability for natural resource management. Where there is tension between conservation and tourism development goals and the inherently complex conflict of interests among stakeholders, conflict resolution is often not a straight forward and uni-directional one. In politically-charged societies where Malaysia is typically one, a compromise between scientific research findings and a wider political consensus looks more practical and implementable. In such cases, good scientific studies which are useful documents for reference must however not be ignored.

References

Aihara, Y., Hosaka, T., Yasuda, M., Hashim, M., & Numata, S. (2016). Mammalian wildlife tourism in South-East Asian tropical rainforests: The case of Endau Rompin National Park, Malaysia. *Journal of Tropical Forest Science, 28*(2), 167–181.

Borrini-Feyerabend, G., & Hill, R. (2013). Governance for the conservation of nature. In G. L. Worboys, M. Lockwood, A. Kothari, S. Feary, & I. Pulsford (Eds.), *Protected area governance and management* (pp. 169–206). ANU Press.

Brockington, D., & Igoe, J. (2006). Eviction for conservation: A global overview. *Conservation and Society, 4*(3), 424–470.

Cárdenas, D. A., Byrd, E. T., & Duffy, L. N. (2015). An exploratory study of community awareness of impacts and agreement to sustainable tourism development principles. *Tourism and Hospitality Research, 15*(4), 254–266.

Carrier, J. G., & Macleod, D. V. L. (2005). Bursting the bubble: The socio-cultural context of ecotourism. *The Journal of the Royal Anthropological Institute, 11*(2), 315–334.

Clayton, A. (2002). Strategies for sustainable tourism development: The role of the concept of carrying capacity. *Social and Economic Studies, 51*(1), 61–98.

Cohn, J. P. (2002). Environmental conflict resolution. *Bioscience, 52*(5), 400–404. https://academic.oup.com/bioscience/article/52/5/400/236102

Dalhberg, A., Rohde, R., & Sandell, K. (2010). National parks and environmental justice: Comparing access rights and ideological legacies in three countries. *Conservation & Society, 8*(3), 209–224.

de Sousa, R. C., Pereira, L. C. C., da Costa, R. M., & Jimenez, J. A. (2014). Tourism carrying capacity on estuarine beaches in the Brazilian Amazon region. *Journal of Coastal Research*, Special Issue No. 70, 545–550.

IUCN. (2021). *About IUCN*. https://www.iucncongress2020.org/about/about-iucn. Dated 3–11 Sept 2021.

IUCN. (2023). *New initiative to support indigenous contributions to biodiversity conservation and climate solutions*. https://www.iucn.org/press-release/202308/new-initiative-support-indigenous-contributions-biodiversity-conservation-and. Dated 09 Aug 2023.

Jaafar, M., Ismail, S., & Rasoolimanesh, S. M. (2015). Perceived social effects of tourism development: A case study of Kinabalu National Park. *Theoretical and Empirical Researches in Urban Management, 10*(2), 5–20.

Kothari, A. (2008). Protected areas and people: The future of the past. *Parks (IUCN), 17*(2), 23–34.

Llausas, A., Vila-Subiros, J., Pueyo-Rosb, J., & Fraguell, R. M. (2019). Carrying capacity as a tourism management strategy in a marine protected area: A political ecology analysis. *Conservation and Society, 17*(4), 366–376.

Mackintosh, B., McDonnell, J. A., & Sprinkle, J. H. (2018). The national parks: Shaping the system. *The George Wright Forum, 35*(2), 1–132.

MountKinabalu.com. (2022). *Sabah parks to increase entrance fees effective 1st January 2023*. https://www.mountkinabalu.com/news/sabah-parks-to-increase-entrance-fees-effective-1st-january-2023

MountKinabalu.com. (2023). *Kinabalu Park*. https://www.mountkinabalu.com/mount-kinabalu/kinabalu-park

Nicholson, D. (2009). Environmental dispute resolution: Theoretical and Indonesian perspectives (Chapter 1). In *Environmental dispute resolution in Indonesia* (pp. 1–47). Brill Stable. https://www.jstor.org/stable/10.1163/j.ctt1w8h1fc.6

O'Connor, S., Pannell, S., & Brockwell, S. (2013). The dynamics of culture and nature in a 'protected' Fataluku landscape. In S. Brockwell, S. O'Connor, & D. Byrne (Eds.), *Transcending the culture–nature divide in cultural heritage* (pp. 203–233). ANU Press.

Seidl, I., & Tisdell, C. (1999). Carrying capacity reconsidered: From Malthus' population theory to cultural carrying capacity. *Ecological Economics, 31*(3), 395–408.

Sievänen, T. (2004). Recreation: Inventory, monitoring and management. *Encyclopedia of Forest Sciences pp.*, 958–964. https://doi.org/10.1016/B0-12-145160-7/00165-4

Sulehan, J., & Gabir, J. (2021). *Conservation tourism on Mantanani Island, Sabah: Impact on the indigenous community*. Institute of Strategic and International Studies. https://www.jstor.org/stable/resrep29683.8

Törn, A., Siikamäki, P., Tolvanen, A., Kauppila, P., & Rämet, J. (2008). Local people, nature conservation, and tourism in northeastern Finland. *Ecology and Society, 13*(1), 8. [online] http://www.ecologyandsociety.org/vol13/iss1/art8/

UNESCO. (2023). *Kinabalu Park*. World Heritage Convention. https://whc.unesco.org/en/list/1012/

Wani, H. A. (2011). Understanding conflict resolution. *Centre for Promoting Ideas, 2011*, 104–111. http://www.ijhssnet.com

Tai-Chee Wong has his BA and MA degrees in Urban and Regional Planning from the University of Paris and earned his PhD in Human Geography from the Australian National University. After teaching 20 years at Nanyang Technological University, Singapore, he retired in 2013. He then worked as Distinguished Professor for two years at Guizhou University of Finance and Economics, China, and as Dean and Professor at the Southern University College, Johor Malaysia, from March 2017 to December 2018, and as Deputy Vice-Chancellor at Berjaya University College, Kuala Lumpur from October 2022 to September 2023. He was a Visiting Professor at the University of Paris (Sorbonne IV) and a Visiting Fellow at Peking University and Tokyo University. His main research interests are in urban and economic issues and, more recently, in Malaysian politics. Besides his 16 self-authored and edited book volumes, he has written over 100 academic articles and published widely in international journals including tourism papers.

Chapter 7
Thai Cultural Tourism Attributes: Emerging Trends and Sustainable Practices

Kashif Hussain, Murali Raman, Mohammad Falahat, and Yasin Ahmed Siddiqui

1 Introduction

The word 'tourism' comprises a broad variety of activities including hiking, camping, vacations, and visiting museums and historical places. The tourism industry is responsible for generating billions of dollars in revenue all across the world every year. The term "cultural tourism" refers to a form of tourism that emphasizes the vibrant cultural heritage of a particular destination. It includes participating in cultural events, going to museums, historical sites, and social gatherings, and getting familiar with local customs. This study attempts to examine not only the physical artifacts and intangible cultural attributes of human society inherited from the past, but includes also in a broader context the inherited natural landscapes. Indeed, the distinction between human-led and nature-based cultural landscape has become increasingly blurred in recent years (Mousavi et al., 2016). In socio-economic sense, both elements have played an equally important role by drawing in tourists in inter-cultural interactions and consumption exchange (Richards, 1996).

Success of cultural tourism is largely dependent on two basic issues: the richness of tangible and intangible cultural assets, and tourism policy coupled with user-friendly approaches and environmentally sustainable measures. Thailand's natural beauty is characterized by its diverse landscapes, which provide opportunities for nature enthusiasts to engage in a broad variety of ways. Breathtaking beaches such as Patong on Phuket, Railay on Krabi, and the islands of Koh Samui and Phi Phi have caught and turned international visitors to join sports events as they are popular destinations for those who like sunbathing and water sports (Henkel et al., 2006).

K. Hussain (✉) · M. Raman · M. Falahat · Y. A. Siddiqui
Asia Pacific University of Technology & Innovation, Kuala Lumpur, Malaysia
e-mail: kashif.hussain@apu.edu.my; murali@apu.edu.my; mohammad.falahat@apu.edu.my

T.-C. Wong et al. (eds.), *Cultural Tourism in the Asia Pacific*,
https://doi.org/10.1007/978-3-031-63459-8_7

Majestic northern Thai cities such as Chiang Mai and Chiang Rai offer a tranquil escape and a wide variety of fantastic hiking opportunities, and these cities are surrounded by green mountains such as Doi Inthanon or other mountains. Enchanting rainforests are another witness of such diverse and natural landscape of Thailand in which stunning lakes, dense forests, and different animals that can be found in Khao Sok National Park make it an ideal destination for ecotourists and explorers. Unique Islands in Thailand is comprised of a large number of archipelago and each of which offers tourists a wide variety of cultural experiences as well as magnificent natural landscapes (Broad, 2003).

Cultural tourism generates great economic impact as it creates job opportunities, attracts foreign currency, and contributes to the overall national economy. To maintain cultural tourism sustainably, preservation and conservation of cultural artefacts is the basis to ensure its enduring survival. Thailand has great potential given its well-known tourist spots, such as Bangkok, Phuket, and Chiang Mai, which offer a captivating blend of stunning scenery and cultural immersion (Berger, 2014). These attractions play a crucial role in boosting the national economy and promoting global interactions with millions of visitors touring Thailand each year.

This chapter explores Thailand as a destination for cultural tourism, analyzing the various attractions that contribute to its distinctive appeal and attractiveness to international travellers. Thailand is an attractive destination because it offers a diverse range of attractions, such as luxurious temples, important historical sites, and compelling natural landscapes that include mountains and beaches (Henkel et al., 2006). This study focuses on the cultural and gastronomic aspects by exploring key tourist sites and rising trends that define the Thai cultural tourism experience. This chapter presents a detailed cultural aspect of the various and constantly changing elements that make Thailand a rich source of cultural discovery. The chapter will finish by summarizing Thailand's remarkable potential as a prominent cultural tourist destination and emphasizes the importance of adopting sustainable practices to ensure a resilient future. The historical evolution of Thai tourism is now discussed.

2 History of Thailand's Cultural Tourism and Sustainable Development

The pre-twentieth century was an era when tourism in Thailand started to document the broad and wide cultural history that spans several historical eras around the nation previously known as Siam (Boonratana, 2010). The rich and vast history of the country was inspired by the civilizations of Indians, Chinese, Khmers and Malays which acted as the driving force influencing the Thai cultural identity. Some of the most magnificent Thai temples and statues such as Wat Phra Kaew and Wat Arun in Bangkok were constructed during the reigns of the Ayutthaya and Rattanakosin dynasties (1351–1767). They have significant historical and cultural values.

During the nineteenth century under the reigns of Kings Rama II and Rama III, a number of Western explorers began their exploratory adventures in Siam. Out of Western curiosity and from their initial encounters with the Thai culture, they were charmed by the Grand Palace and Wat Phra Kaew which later became the epitome of the Thai cultural heritage that has a strong appeal to international tourists till today (Cohen, 1987). From the end of the nineteenth century to the early twentieth century, it was a period characterized by modernization and cultural preservation. King Rama V of Thailand was known to have made efforts to modernize the country while also preserving and displaying the country's rich cultural legacy (Cohen, 1982). One of the most important events was the king's effort in preserving the cultural relics by setting up the Department of Fine Arts in 1910, thus providing major protection to museums and historical monuments.

The Thai tourism industry experienced a serious setback during World War II and the immediate post-war period. From the 1950s however, a few cultural tourism campaigns were launched to highlight the Thai distinctive traditions, ceremonial events and architectural wonders to the international community (Hvenegaard & Dearden, 1998). This had brought considerable success, and during the three decades that followed (1960s to the 1980s) Thailand's tourism witnessed a "Boom" with a considerable increase in the number of international tourists. Among the touristic features that had captured the interests of tourists were significantly linked to the country's enormous historical and cultural legacies such as temples, and authentic traditional performances, floating markets, among others (Berger, 2014).

In support of Thailand's industrial take-off and the promotion of tourism development, the 30 years from the 1980s to 2000s saw substantial investment in transportation infrastructure, notably highways and airports and this had provided better physical access to cultural sites. Beside Bangkok, the city of Chiang Mai and the island of Phuket were the two most important beneficiaries for being key cultural centres. From the 1980s, Thai's gastronomy tourism has begun to supplement its cultural tourism and has emerged as one of the unique cooking classifications well received by tourists across the globe. Food tasting trips to local Thai markets, and participation in culinary festivals have become equally a common part of cultural tourism programs (Cohen, 1982; Cohen & Avieli, 2004).

From the 2010s onwards, as part of the global development trend, ecotourism and adventure travel tours have witnessed an increase in the number of people taking part. Thailand's tropical and sub-tropical natural landscapes which include coasts, archipelagos, and forests have attracted a new group of cultural tourists who are looking for immersive experiences that extend beyond metropolitan zones (Hvenegaard & Dearden, 1998). In this relatively new phenomenon, it is critical to associate the explorations of natural adventures to the notion of sustainable cultural tourism which is the key concern of the twenty-first century where a significant emphasis is placed on the promotion of environmentally responsible cultural tourism (Boonratana, 2010). In order to fulfill the goal of achieving a harmonious balance between the protection of cultural assets and the promotion of tourism, we will examine the measures which have been undertaken in the preservation of Sukhothai and Ayutthaya classified as world heritage sites and preservation of Pattaya and Phuket as charming Thai beaches.

2.1 Sustainable Cultural Tourism

Sustainable cultural tourism is an ethical and reasonable approach towards tourism in promoting the cultural heritage of a particular destination (Sangkaew & Zhu, 2022). This is the kind of tourism in which tourism seeks a balanced approach between the needs of the tourists and the long-term viability of the well-being of the host community as a cultural heritage. Hence, the preservation of the historical heritage and beaches of Thailand are particularly discussed below.

2.1.1 Preservation of World Heritage Sites: Sukhothai and Ayutthaya

Among the world's most famous heritage sites, Thailand's Sukhothai and Ayutthaya are two symbolic legacies which have left behind permanent historical footprints. Under the reign of the third king of the Phra Ruang Dynasty, King Ram Khamhaeng who created Thai script to give the nation its identity, Sukhothai as the capital city became famous for the Thai culture and economic prosperity. Phra Ruang dynasty (1238–1438) lasted about 200 years, and the golden era was the initial period till AD 1300 whereby many glorious monuments were built. The kingdom was invaded and annexed by Ayutthaya, a neighboring Thai state in 1438 (Inthajakra et al., 2016).

The Ayutthaya Kingdom had its capital built to the south of Sukhothai and 86 km north of the present day Bangkok, known then as Phra Nakhon Si Ayutthaya. After annexing Sukhothai in 1438, Ayutthaya city became a glorified centre as one of the largest cities in Southeast Asia. The Ayutthaya Kingdom lasted 417 years (1350–1767), and it enjoyed a flourishing culture characterized by Buddhist art, and established good trade and diplomatic relations with neighbouring states, as well as China, Japan and even Europe, leaving an indelible mark on Thai culture (Smith, 2017). Ayutthaya's strategic location along the lower reaches of the Chao Phraya River made it a trade edge for the kingdom (Baker & Phongpaichit, 2017).

Both Sukhothai and Ayutthaya cities have been elected by UNESCO as World Heritage sites, which is a good recognition of their enriched preserved Thai cultural heritage. The historic temple of Sukhothai has been made a UNESCO site since 1991 (Patterson, 2012). With subsequent conservation efforts, a strong governmental engagement has been dedicated in maintaining this historical site where valuable elements of archaeology, monumental statues are meticulously cared for. Not only the historic sites of Ayutthaya and Sukhothai are safeguarded in their building framework, constant repairing and restoration of the structural defects have also been looked into thoroughly under the conservation programs (Saipradist & Staiff, 2008). Public tourism management agencies have been conducting sustainable practices to protect fragile structures and they also educate tourists to respect sites for their sustainability values.

2.1.2 Preservation of Thai Beaches: Pattaya and Phuket

Preserving the ecological integrity of the beaches, their biodiversity, cultural significance and assuring management effectiveness are critical in keeping their authentic natural identity and features. In Thailand, Pattaya and Phuket are among the two most popular beaches, famous for their natural scenery as popular tourist spots. To help sustain their beauty, the Thai authorities have set a high international standard to preserve them as environmentally friendly and natural beaches (Choeirod & Chanthao, 2023).

To make both beaches a better place for tourists, the Thai government has followed a 'go green' master plan. With local community involvement, the government has launched green programs and issued licenses to encourage agencies and visitors to preserve the natural beauty of the beaches. The Eastern Economic Corridor (EEC) Project is one which aims to maintain the overall attractiveness of the beaches. In this regard, Green Certification Programs being implemented are in recognition of the need as a sustainability practice to keep an eco-friendly environment for local and international tourism (Sangkaew & Zhu, 2022). Using the Waste Management Initiatives to control disposals which is an eco-friendly tourism practice, beaches, nearby green spaces and parks are maintained as far as possible at a high aesthetic level. Other sustainable measures adopted include more efficient energy accommodation by energy recycling, and reducing traffic congestion to facilitate ecological travelling (Fuchs & Kris, 2021).

2.1.3 Gradual Move Towards Sustainable Cultural Tourism

Thailand is committed to ecotourism and conserving the environment, which is in line with the global eco-friendly initiatives to preserve natural resources and encourage responsible travel. The government of Thailand has taken different initiatives via the implementation of eco-friendly programs such as the reduction of wastes, energy saving and the use of sustainable transportation options (Hvenegaard & Dearden, 1998). In the tourism industry, Thailand adopted in 2018 sustainable practices with a comprehensive act for the conservation of natural resources in which coral reefs, national parks, and wildlife habitats are preserved to ensure that they will continue to exist in good natural state for generations to come (MNRE, 2019).

Thailand is currently developing appropriate legislation and guidelines to address the issues of responsible tourism and sustainability. In order to achieve this objective, the Thai government felt it is imperative that they create and rigorously comply with ecological and ethical criteria for the tourism sector. In this objective, they have implemented comprehensive planning and zoning measures to regulate tourist activities and the growth of infrastructure (Boonratana, 2010). In order to align standards with global best practices, they have engaged in partnerships with international organizations and authorities in sustainable tourism by which public awareness campaigns have been launched to serve as a means to highlight the significance of ethical travel to both visitors and companies. Among these efforts, establishing a sustainable tourist framework to encompass waste management systems, to set the maximum number of visitors to environmentally sensitive places has been implemented.

In addition to Thailand's commitment to sustainable tourism practices, community engagement in which local governing bodies and small communities have been empowered at the grassroots level. This initiative has worked proactively on issues pertaining to localized environmental hazards (Boonratana, 2010). Thailand's cultural legacy, diverse landscapes, friendly peoples, and commitment to ecotourism are some of the potential unique selling propositions that the country may offer foreign visitors (Williams, 2002).

3 Diverse Attractions in Thai Cultural Tourism

Having highlighted public sustainability measures and practices in promoting the cultural tourism of Thailand, we now move on to illustrate in greater details some selected attractions possessing symbolic representative values in the country, categorized as given below.

3.1 Religious Temples, Museums and Architecture

Wat Pho (Temple of the Reclining Buddha) is one of the largest and oldest temples in Bangkok and is the most splendid one. The prominent attraction portrays a colossal reclining statue of the Buddha, together with a vast assortment of Buddha pictures and excellent paintings. Wat Arun (Temple of Dawn) is a religious place which is situated on a high position alongside the banks of the Chao Phraya River. The central peak of the tower is intricately ornamented with vibrant ceramics and seashells. The temple offers awe-inspiring panoramas over the encompassing region encompassing the river and the urban centre (Hvenegaard & Dearden, 1998).

Bangkok's National Museum holds a huge collection of objects that richly show the historical and cultural aspects of Thailand. Relics encompass ceramics, religious artifacts, and traditional artwork that are among the famous remnants of the National Museum. Ayutthaya Historical Park is an open museum famous for Thailand's rich cultural and historical heritage and it shows the ancient city form of Ayutthaya which is classified as a UNESCO World Heritage site (Cohen, 1982). The city is decorated with several monuments and temples with archaeological treasures. The Park has artefacts that provide proof of the exceptional architectural skill that was widespread throughout that time.

Sukhothai Historical Park is another UNESCO World Heritage Site with well-preserved temple and monument ruins from the Sukhothai Kingdom (Hvenegaard & Dearden, 1998). The Grand Palace is a prominent structure located in the heart of Bangkok; it showcases the originality and architectural design of old Thailand.

The premises consist of the Emerald Buddha Temple, as well as several places and pavilions embellished with intricate patterns. Chiang Mai Old City is the ancient city of Chiang Mai embellished with revered structures including Wat Chedi Luang and Wat Phra Singh fortified with antique walls and gateways (Cohen, 1987).

3.2 *Traditional Visual and Performing Arts and Music*

Every year, there are Thai cultural festivals and events which are celebrated with traditional rituals, musical and dance performances. The traditional Thai masked dance, called Khon dance, is performed in important events. It is a Thai adaptation from the story of the heroic figure Ramayana, originally written in an Indian language (Broad, 2003). Khon dance is characterized by the use of intricate masks, luxurious costumes, and precise choreography.

Thai Boxing Shows (Muay Thai) are world famous for its traditional martial art which demonstrates the physical strength and ferocity of fighters during performance where fighters participants wear unique costumes in their display (Boonratana, 2010). Loy Krathong Festival also known as the Festival of Lights is a national holiday where participants are encouraged to practice a relaxation technique by placing their "worries" under the decorative containers called krathongs floating on water. Another important celebration is the Thai New Year, known as Songkran, held annually in the hot month of April whereby water plays an important role and it is used to pour on people to help cool off the heat. This practice has aroused disputes with tourists in the cities these days despite its longstanding tradition as a highly renowned and warmly welcomed festival (Broad, 2003; Biggs et al., 2012).

Musical performance is an indispensable component of Thai festivals and Thai dance and drama. The Thai traditional music is distinguished by the use of several instruments, including hammered dulcimers, saw duangs, xylophones, and khims. Live entertainment is sometimes present during classical ceremonies and traditional festivals. Thai theatre, known as "Likay," is a captivating cultural performance that combines acting, singing, and dance to reflect Thai art and music (Henkel et al., 2006).

3.3 *Traditional Artistic Crafts and Thai Cuisine*

Thai silk and textile making is a rich tradition of ancient weaving techniques that have been passed down through centuries and is a significant component in Thailand's impressive silk production. There are charming silk villages that attract travellers to visit and witness the sophisticated production process. Sukhothai area is known for the Thai traditional artistic pottery and ceramics manufacturing where exquisite products are intricately designed and made by talented craftsmen (Williams, 2002).

Rich Thai culinary tradition has gained worldwide recognition due to its diverse menu, lively tastes, and strong spices and these have become a reason of visit to Thailand. Besides, Thailand's highly dynamic and street foods and small markets offer a wide array of culinary options such as satay, pad Thai, som tum (green papaya salad), and mango sticky rice are commonly found in the street markets in Bangkok and other large cities (Cohen & Avieli, 2004). There are a large number of culinary schools, offering cooking classes in tourist locations for visitors to learn the art of making traditional Thai food and culinary events are also widely organized (Cohen, 1987).

4 Emerging Trends in Cultural Tourism

The current developments in cultural tourism are centred around ecotourism and adventure travel. There is a rising attraction towards Thailand's beaches, islands, and national parks. The following are the prevailing travel preferences:

4.1 Rise of Ecotourism and Adventure Travel

Ecotourism places emphasis on sustainable development of cultural tourism with the goals of protecting the natural areas and it has raised public awareness in promoting sustainable practices. One of the key objectives is to minimize visitors' ecological footprint while improving the local community's quality of life. Popular ecotourism destinations are not rare due to the country's diverse range of habitats, including highland, tropical, and underwater areas. Khao Sok and Doi Inthanon national parks as well as islands and beaches provide environmentally-friendly trips with their natural diversity (Hvenegaard & Dearden, 1998). Sites like these provides also a wider range of adventure activities such as rock climbing, ziplining, hiking, and diving.

An increasing number of travellers to Thailand have chosen community engagement to get themselves immersed with local culture and practices. The current trend has indicated a shift towards holidays that stress both thrilling experiences and environmental stewardship.

One element worthy of mentioning is the hospitality and warmth of the Thai people towards international tourists. Placing high value on courtesy and respect to strangers has contributed to the maintenance of positive relationships between locals and tourists. Among the traditional Thai greetings, use of the *wai* (bow with two folding palms) is one way to bring a more personal touch to interpersonal inter actions that convey respect and kindness (Broad, 2003).

5 Challenges and Future Infrastructure Development

5.1 Overcoming Challenges Related to Over Tourism

An issue of significant magnitude in Thailand pertains to the volume of tourists, particularly in renowned tourist destinations such as Chiang Mai, Phuket, and Bangkok. Consequences of overcrowding, environmental hazards, depletion of local resources, and disruption to local inhabitants are a few examples. In addressing these problems, Thailand has used several strategies like local community engagement, creating awareness and promoting educational programs to counter the adverse effects (Boonratana, 2010). Other tourism programs seeking to decentralize tourists to less visited destinations, provide incentives for off-peak travel schedules and improvement of management techniques have also been undertaken.

5.1.1 Balancing Economic Benefits with Cultural and Environmental Preservation

Over tourism may yield both beneficial and harmful consequences. Therefore, striking a balance in this aspect to achieve an equilibrium level is critical. The impact by tourism on Thailand's varied ecosystems and cultural legacy must be balanced on the basis of economic benefits and environmental conservation. Tourism that is not regulated leads to the commodification of local cultures, compromises in authenticity, and causes harm to the environment. Thailand particularly values responsible tourism that demonstrates respect for local cultures and the environment in which their local communities are urged to engage in and enjoy the advantages of tourism through community-based initiatives (Boonratana, 2010). Corporate eco-certification, responsible tourist efforts, and cultural preservation programs are all primary elements of the solution. In order to achieve this balance, Thailand's government, communities, and corporate must cooperate in safeguarding the nation's cultural and natural assets as the economy expands.

Thailand has now acknowledged the issues arising from over-tourism, the need of enacting good legislation to encourage sustainable tourism and the need to achieve a harmonious balance between economic advantages and regulations. Thailand intends to tackle these challenges by thorough planning, collaborative efforts, and a commitment to responsible practices. The objective is to establish a future in which tourism and the regulation of Thailand's cultural and natural assets can coexist in a peaceful manner.

5.1.2 Investments in Cultural Infrastructure and Preservation Projects

Thailand has given priority to investments in cultural infrastructure and preservation projects as part of its long-term tourism aspirations. Government agencies have been dedicated the role with financial support to save cultural items, temples, and historical places (Biggs et al., 2012). As said earlier, the Grand Palace in Bangkok, the Ayutthaya Historical Park, and the Sukhothai Historical Park all fall under this classification of restored landmarks. In order to enhance tourists' understanding of Thailand's abundant cultural legacy, investment initiatives encompass the construction of more museums, cultural hubs, and more educational establishments.

Thailand is utilizing technology to develop innovative and user-friendly solutions that improve the travel experience while also meeting environmental standards. In doing so, Thailand has embraced technological innovations to improve tourism services and facilities historic and cultural sites to meet the demands of tech-savvy tourists who expect augmented and virtual reality experiences. Travellers may utilize digital platforms and applications that offer interactive tours, translation services, and seamless navigation. Furthermore, the tourism industry may contribute to environmental preservation by using sustainable technologies, such as energy-efficient systems and eco-friendly transportation options (Berger, 2014).

Another effort that Thailand has dedicated is the fostering collaborations between the public and business sectors to ensure a complete and unified approach to tourist development. Public-private partnerships play a vital role in funding and managing extensive tourist projects, such as infrastructure development, marketing campaigns, and destination administration. The objective of this joint effort is to capitalize on the expert knowledge of governmental supports, district societies, and commercial firms towards establishing a tourist sector that fosters local economies and communities via shared accountability. Collaborative projects encourage sustainable practices, cultural authenticity, and responsible tourism.

Thailand's future infrastructure and development objectives rely on the crucial factors of public-private partnerships, investments in cultural preservation, and technological breakthroughs to enhance visitor services. These programs seek to establish Thailand as a forward-thinking and environmentally conscious tourist destination that can accommodate the changing demands of tourists while preserving the country's abundant cultural heritage and breath-taking natural beauty of the country.

6 Conclusion

Thailand is a multifaceted treasure that offers an unmatched blend of cultural assets, varied landscapes, welcoming hospitality, and ecotourism. Exploring historical places such as Ayutthaya and magnificent temples like Wat Pho showcases Thailand's prestigious cultural legacy. The natural beauty of the area is enhanced by the magnetism of its pristine beaches, lush mountains, and thriving rainforests. Furthermore, Thai people extend a warm and grand reception to tourists enhancing their experience with captivating traditional festivals and arts. The government is actively devising strategies to address issues such as over-tourism, while also aiming to achieve a harmonious coexistence between economic advantages and cultural conservation, as it plans for the future.

Thailand's potential in cultural tourism remains unmatched as visitors from all over the world are fascinated by its abundant historical significance, varied cultural landmarks, and welcoming people. Thailand has been a prominent global leader in cultural tourism due to its technical progress, preservation efforts, and investments in cultural infrastructure. The country's commitment to accountable and enduring tourism is seen in the cooperation between the governmental and corporate spheres. Thailand's capacity to progress while preserving its distinctiveness ensures its ongoing appeal to travellers in search of genuine and gratifying encounters indicating a promising future for the nation as a cultural tourism front-runner.

Thailand is witnessing an increasing trend towards adopting ethical and eco-friendly practices in the tourism sector. Collaborative efforts are required to address the delicate balance between economic expansion and cultural preservation, as well as the risks associated with over-tourism. The government's commitment to environmentally-friendly programs, collaborations, and technological advancements demonstrates its sincere efforts to develop a sustainable tourism business that benefits both the environment and the local community, while enjoying enormous economical benefits. Responsible travel encourages both tourists and stakeholders to promote cultural immersion, welfare, and a real appreciation for the unique characteristics of Thailand. If Thailand adheres to these guidelines, it may maintain its priceless historical heritage for an extended period while being a sought-after cultural tourism destination.

References

Baker, C., & Phongpaichit, P. (2017). *A history of Ayutthaya*. Cambridge University Press.

Berger, A. A. (2014). *Thailand tourism*. Routledge.

Biggs, D., Hall, C. M., & Stoeckl, N. (2012). The resilience of formal and informal tourism enterprises to disasters: Reef tourism in Phuket, Thailand. *Journal of Sustainable Tourism, 20*(5), 645–665.

Boonratana, R. (2010). Community-based tourism in Thailand: The need and justification for an operational definition. *Kasetsart Journal of Social Sciences, 31*(2), 280–289.

Broad, S. (2003). Living the Thai life - A case study of volunteer tourism at the gibbon rehabilitation project, Thailand. *Tourism Recreation Research, 28*(3), 63–72.

Choeirod, R., & Chanthao, R. (2023). Linguistic landscape at Patong, Phuket, Thailand: Language perception on public signs. *NeuroQuantology, 21*(1), 250–261.

Cohen, E. (1982). Marginal paradises: Bungalow tourism on the islands of southern Thailand. *Annals of Tourism Research, 9*(2), 189–228.

Cohen, E. (1987). "Alternative tourism"—A critique. *Tourism Recreation Research, 12*(2), 13–18.

Cohen, E., & Avieli, N. (2004). Food in tourism: Attraction and impediment. *Annals of Tourism Research, 31*(4), 755–778.

Fuchs, K., & Kris, S. (2021). The status quo of sustainable tourism development in Phuket: A qualitative assessment. *Journal of Environmental Management and Tourism, 1*(49), 167–172.

Henkel, R., Henkel, P., Agrusa, W., Agrusa, J., & Tanner, J. (2006). Thailand as a tourist destination: Perceptions of international visitors and Thai residents. *Asia Pacific Journal of Tourism Research, 11*(3), 269–287.

Hvenegaard, G. T., & Dearden, P. (1998). Ecotourism versus tourism in a Thai National Park. *Annals of Tourism Research, 25*(3), 700–720.

Inthajakra, L., Prachyapruit, A., & Chantavanich, S. (2016). The emergence of communication intellectual history in Sukhothai and Ayutthaya kingdom of Thailand. *Social Science Asia, 2*(4), 32–41.

MNRE (Ministry of Natural Resources and Environment). (2019). Enhancement and Conservation of National Environmental Quality Act (No.2) B.E.2561 (2018) https://www.mnre.go.th/en/information/list/1387 Added.

Mousavi, S. S., Doratli, N., Mousavi, S. N., & Moradiahari, F. (2016). *Defining cultural tourism*. International conference on civil, architecture and sustainable development held in London, December 1-2, 2016. https://doi.org/10.15242/IICBE.DIR1216411.

Patterson, J. L. (2012). Ostentation and invisibility: The Phra Phutthabat and royal pilgrimage in late Ayutthaya. *Artibus Asiae, 72*(1), 159–192.

Richards, G. (1996). *Cultural tourism in Europe*. CAB International.

Saipradist, A., & Staiff, R. (2008). Crossing the cultural divide: Western visitors and interpretation at Ayutthaya world heritage site, Thailand. *Journal of Heritage Tourism, 2*(3), 211–224.

Sangkaew, N., & Zhu, H. (2022). Understanding tourists' experiences at local markets in Phuket: An analysis of TripAdvisor reviews. *Journal of Quality Assurance in Hospitality & Tourism, 23*(1), 89–114.

Smith, R. (2017). *The kings of Ayutthaya: A creative retelling of Siamese history*. Silkworm Books.

Williams, S. W. (2002). *Tourism geography*. Routledge.

Kashif Hussain has more than 22 years of professional experience demonstrating educational leadership and research excellence. He serves as Professor and Head of the School of Global Hospitality and Tourism at the Asia Pacific University of Technology and Innovation in Malaysia. He holds a PhD in Educational Administration and Supervision from Near East University and an MSc in Tourism Management form Eastern Mediterranean University, North Cyprus (Turkey). In addition to being a professor of hospitality and tourism studies, he is the author of more than 120 books, journal articles, book chapters, and international conference proceedings.

Murali Raman is a Rhodes Scholar and Fulbright Fellow. His academic credentials include a PhD from SISAT, Claremont, USA, an MBA from Imperial College, London, and an MSc in Human Resources from the London School of Economics, UK. He is affiliated with Stanford's Technology Venture Program as a Faculty Fellow where he continues to discuss issues surrounding creating a vibrant entrepreneurship ecosystem and the application of innovative thinking via Design-Principles in Malaysia. He is a Stanford-certified Design Thinker. He has published more than 120 papers in international journals, conferences, and book chapters, in management information systems related areas.

Mohammad Falahat is a distinguished academic and industry expert, currently a Professor at the Asia Pacific University of Technology & Innovation (APU) and Adjunct Professor at Universiti Tunku Abdul Rahman (UTAR), Malaysia. He has achieved notable recognition, including the USM Gold Medal Award for his doctorate and significant contributions to high-impact journals in his fields of expertise. In the realm of scholarly publishing, Falahat published in high-impact journals such as Corporate Social Responsibility and Environmental Management (SSCI: 8.74), Technological Forecasting and Social Change (SSCI:5.8), Journal of International Management (SSCI:3.8), and International Marketing Review (SSCI: 2.9) to name a few.

Yasin Ahmed Siddiqui is currently pursuing a PhD in Management in Asia Pacific University of Technology & Innovation, Malaysia. Earlier, he held a leadership role in the Entrepreneurship program at Muhammad Ali Jinnah University in Pakistan where he lectured and founded several commercial ventures. He is skilful in teaching to maintain students' academic standards. He has attended several conferences to gain exposure in real-world research, and has participated in a number of research projects.

Chapter 8
Cultural Districts and Tourism in Hong Kong: A Case Study of West Kowloon Cultural District

Benny Lim

1 Introduction

Hong Kong is a vibrant cosmopolitan city and is often regarded as one of the most popular tourist destinations in the world. The city is attractive in many ways, from its iconic skyline, diverse culinary scene, shoppers' paradise, to its rich cultural heritage. However, Hong Kong's tourism sector has faced significant challenges since 2019, which were caused by two independent major events, i.e., the Anti-extradition Bill Protests of 2019 and 2020, as well as the COVID-19 pandemic since early 2020. The large-scale demonstrations in 2019 led to a decrease of close to 10 million tourists, down from 65.1 million visitors in 2018 to 55.9 million in 2019 (To & Lee, 2023). The pandemic further exacerbated the situation. The city adopted some of the world's strictest COVID-19 measures, which resulted in the complete shutdown of the territory for most of 2020 and 2021. In 2022, the visitors' numbers were a mere 0.6 million (Hong Kong Tourism Board, 2023), a far cry compared to the statistics from years before. Needless to say, there was a steep decline in tourism's contribution to Hong Kong's GDP between 2018 and 2022.

Nevertheless, amidst these adversities in Hong Kong, the West Kowloon Cultural District (WKCD) has gained significant traction amongst the Hong Kong people in these recent years. The idea of WKCD was first mooted in 1998, a year after Hong Kong's return to China, by then Chief Executive Tung Chee-Hwa. The cultural district is, by far, the most ambitious cultural investment made by the Hong Kong government, with the mission to promote arts and culture, as well as the long-term development of the city's cultural and creative industries. The unique position of Hong Kong as a special administrative region of China also propels WKCD's role

B. Lim (✉)
The Chinese University of Hong Kong, Hong Kong, Hong Kong
e-mail: bennylim@cuhk.edu.hk

T.-C. Wong et al. (eds.), *Cultural Tourism in the Asia Pacific*,
https://doi.org/10.1007/978-3-031-63459-8_8

in reinforcing the city's image where East meets West, traditional interweaves with contemporary, and local culture engages with global trends. It has not always been a bed of roses for the cultural district. During the initial stages, there were public outcries about the district being yet another commercial property project. Throughout the construction period, there were debates about the delays in the construction, the excessive government spendings, as well as the high-profile resignations (Lim, 2018). Despite this, the Hong Kong government persisted with this mega project, reflecting high levels of commitment and resilience.

The district saw the launch of four major cultural institutions over the span of 3 years. They include, Xiqu Centre in early 2019, Freespace in mid-2019, followed by M+ Museum (M+) and the Hong Kong Palace Museum (HKPM) in November 2021 and July 2022 respectively. It is worth noting that both museums inaugurated amid stringent COVID-19 restrictions and lockdowns. By the end of 2022, Hong Kong lifted all COVID-19 measures, including border restrictions, and in early 2023, the city reopened its borders with Mainland China after almost 3 years. Undoubtedly, these events shed light into a new era of Hong Kong's tourism and provides a great opportunity for WKCD to further its reach. This chapter, therefore, seeks to explore the strategies that drive WKCD as a cultural destination, and in so doing, debates the potential of cultural districts in boosting tourism in Hong Kong.

2 Cultural Tourism in Hong Kong

Cultural tourism is a complex concept (Duhme, 2012), and its definition has evolved over the years (Richards, 2001). Within the context of tourism studies, cultural tourism is widely understood as a form of special interest tourism (Ramchander, 2007). One widely cited definition of cultural tourism refers to the movement of individuals towards cultural attractions, driven by a quest for new knowledge, information, experiences, as well as the fulfillment of cultural needs (Richards, 2021). Cultural attractions can range from monuments, museums, and galleries to cultural festivals, heritage, and traditions unique to certain locales (Smith et al., 2010). WKCD, for one, makes a great cultural attraction with its mix of beautifully designed museums, performing arts, and outdoor spaces, and they collectively present a range of artistic and cultural works, from exhibitions, performances, screenings, to festivals.

Often, cultural tourists are motivated to seek genuine and authentic experiences through immersing themselves into the lifestyles and rituals of the cultural attractions/destinations (Domínguez-Quintero et al., 2020). The idea of 'authenticity' in cultural tourism, while crucial, is also a highly subjective and debatable term as the preservation of authenticity is often showcased alongside purposefully curated experiences for tourism. Cultural tourism is beyond the mere showcase of cultural artifacts and cultural attractions/destinations seek to generate memorable experiences for visitors through well-curated orchestration of elements such as space, time, and atmosphere (Chhabra, 2022). The Experience Economy model, first conceptualized by Pine and Gilmore (1999), offers insights into this phenomenon.

Pine and Gilmore argue that businesses create value by staging memorable experiences around their goods and services. Nevertheless, these experiences are not singular. It is important for cultural attractions/destinations to offer a varying range of experiences, so as to cater to different types of cultural tourists, some of whom are more active, while others, passive.

It is important to note that there are different types of cultural tourists, suggesting that cultural tourism is a diverse concept. Purposeful cultural tourists have strong interests in experiencing the locale's cultural fabric and heritage, so much so that these interests probably propelled the travel intent in the first place. Casual cultural tourists visit cultural attractions/destinations but such activities are usually carried out without a pre-planned intent (McKercher & Du Cros, 2003). Furthermore, there are also 'accidental' cultural tourists, who may stumble upon cultural experiences without any prior planning or expectations (Richards, 2021). Moreover, not all cultural tourists share the same motivations. Technology has enabled cultural institutions to offer innovative, interactive experiences to visitors, further enriching the cultural tourism offering (Ponsignon & Derbaix, 2020). With the advent of digital technologies and social media platforms, some tourists visit specific cultural attractions for the purpose of generating content (and attention) on their social media platforms (Large, 2019).

Historically, Hong Kong was often regarded as a cultural desert, due to the lack of cultural amenities and attractions in the early- and mid-twentieth century. As a British colony, Hong Kong was often known for its East-West eclecticism, presenting a perceived lack of traditional forms of Chinese culture (Karvelyte, 2018). Being a cultural desert does not mean that Hong Kong has no cultural identity. The city's culture was deeply ingrained in the lives of everyday people, or the 'lived' culture, which was manifested through the everyday street life, the gastronomic offerings, and shopping malls (Ng, 2009). Yet, this side of Hong Kong's culture was overshadowed by the city's reputation as a global financial center. Since the handover in 1997, the government has taken active steps to transform Hong Kong from a 'desert' to a vibrant cultural hub (Karvelyte, 2018), which included the conception of WKCD.

Existing literature suggests that Hong Kong has a diverse range of cultural attractions, from performing arts and cultural exhibitions (Ho & McKercher, 2004), traditional Chinese festivals and rituals (Mackellar, 2014; Ho & McKercher, 2004), gastronomic experiences (Chiu & Huang, 2022; Ho & McKercher, 2004), and heritage architectures and trails (Barber, 2019; Ho & McKercher, 2004). The city's colonial past, coupled with its Chinese heritage, present a unique cultural landscape that has the potential to attract and engage tourists. Given its unique blend of East and West cultures, Hong Kong is also an attractive and accessible destination for non-Asian cultural tourists visiting Asia for the first time (Nguyen & Cheung, 2014). Nevertheless, Hong Kong has yet to be perceived as a primary destination for cultural tourism. Most of the city's tourists are either accidental or casual cultural tourists. In other words, the city has yet to fully leverage its cultural resources for tourism (Tam, 2023).

3 Cultural Districts and Their Values

3.1 Typologies of Cultural Districts

The development of cultural districts is often associated with the growing emphasis of arts and culture in city and urban development (Arnaboldi & Spiller, 2011; Markusen & Gadwa, 2010). A cultural district refers to the geographic concentration of cultural activities and facilities and is closely related to the concept of industrial clustering (Noonan & Breznitz, 2013). Businesses within an industrial cluster can achieve economies of scale and cut back on transportation expenses. The cluster also encourages the concentration of skilled workers and fosters the spirit of both formal and informal mentorship between them (Puig & Marques, 2011). In the context of a cultural district, the clustering allows resources to band together and generate a milieu that stimulates creativity, innovation, and collaborations. This is echoed in cultural economist Walter Santagata's definition of a cultural district as "a social and economic experience at the confluence of two phenomena: that of localization... and that of the idiosyncratic nature of culture and cultural goods" (Santagata, 2011: 147).

A cultural district is often recognized as a cultural landmark, insofar as it is a highly visible landscape within a locale. Nonetheless, cultural districts and cultural landmarks are separate entities. Cultural landmarks are often recognized for their cultural, historical, or architectural significance. They could be standalone iconic buildings/structures, and their significance often outweighs that of their locale. On the other hand, cultural districts tend to focus on the distinctiveness of the locale and promote both immersion and consumption of multiple cultural experiences (Nuccio & Ponzini, 2017). A cultural landmark, such as a performing arts center, could anchor a cultural district, attracting tourists to visit the landmark whilst also exploring the uniqueness of the district (Webb, 2017). Collectively, both entities play a crucial role in enhancing cultural tourism.

Scholars and research organizations have presented different typologies of cultural districts, each with its own distinct characteristics and purpose. Categorizations, such as heritage districts, creative districts, theater districts, are often used interchangeably with cultural districts. Generally, cultural districts are either producer-oriented, consumer-oriented, or both. Santagata (2002) categorized cultural districts into four types, which include industrial cultural districts that focus on the production and creation of (often) related cultural goods and services; institutional cultural districts, represented by a community's collective property rights in cultural production; museum cultural districts, which involve a concentration of museums as well as historic and heritage sites; and metropolitan cultural districts, which sees the development of cultural organizations/institutions, related cultural services alongside other amenities and services such as cafes and restaurants within close proximity.

Sacco et al. (2008) developed the system-wide cultural districts model, which expanded on Santagata's categorizations. In system-wide cultural districts, culture acts as a lever for all the production sectors, typically in the post-industrial contexts, and often span across multiple geographical areas. A close example of a system-wide cultural district is the Bras Basah. Bugis arts and heritage district, a 95-hectare area in Singapore with a concentration of museums, art centers, art university, monuments, and diverse ethnic communities, alongside many shopping malls and lifestyle providers. These landmarks and communities brought together through a series cultural activities and festivals. Simply put, the precinct represents a vibrant part of Singapore, where arts and creativity take lead.

3.2 Values of Cultural Districts

Cultural districts provide significant cultural value through improving the communities they serve and shaping the lives of their residents. One of the benefits that cultural districts provide to residents and local communities is the development of cultural capital. The term 'cultural capital' here refers to the collection of shared knowledge, competencies, and skills that arise from the exposure to, and participation in, cultural activities (Mihelj et al., 2019). Cultural districts enrich this cultural capital by providing a diverse range of cultural and artistic experiences such as art exhibitions, theater performances, concerts, and workshops, allowing individuals to broaden their cultural quotient, increase their understanding and appreciation of various art forms, and engage in creative expressions. Cultural amenities in these districts also offer vibrant public spaces that encourage social interaction and community engagement. All in all, the creative atmosphere brought about by cultural districts improves social and cultural capital as well as the well-being and quality of life of the communities (Santagata, 2006).

Cultural districts also play a pivotal role in developing the local arts and creative sectors in general. They provide artists and creatives, especially emerging ones, with spaces to create and showcase their works, and in so doing, help them to generate livelihoods from their practice. Moreover, through forging partnerships with cultural districts, non-profit arts and cultural organizations can increase their visibility and gain access to wider audiences. By clustering artists, creatives, and organizations together, cultural districts encourage networking, collaboration, and incubation of new creative expressions, which bring about a more diverse range of cultural presentations to the communities (Borrup, 2013). Partnering with local artists and organizations enhance the vibrancy of the cultural districts, contributing to their uniqueness and up their potential to attract cultural tourists.

Cultural districts generate substantial economic value in the communities and locales they serve. For one, cultural districts are important sources of job creation (Santagata, 2006). They hire artists, administrators, and technical professionals, and at the same time, provide entrepreneurial activities for a large pool of freelance creatives. With a diverse range of cultural offerings, cultural districts bring about direct economic impacts through ticket sales and other related merchandise.

Through empowering arts, culture, and creativity for the communities, cultural districts revitalize the neighborhood/city through creative place-making (Noonan, 2013). This leads to an enhanced brand image of the locale, differentiating it from others and increasing its competitive edge. A creative neighborhood/city is bound to become key destinations for both domestic and international tourism, which generates further direct and indirect economic impacts. Tourists spend on accommodation, transport, food, and a range of other retail outlets and services from within the district as well as its surroundings. The vibrancy of economic activities also brings about increased job opportunities for local businesses nearby. On the same note, property prices in and around cultural districts are likely to go up (ibid 2013). Nevertheless, these positive economic impacts do not materialize immediately and automatically. A successful cultural district requires good strategic management alongside investment of capital and other resources.

While cultural districts are known to bring about economic impacts and contributions to the city/nation's GDP, they also bring about several displacements. The increase in tourists, if not well-managed, leads to the commercialization of culture. The original mission and vision of the cultural district may be displaced to cater to the needs and taste of the tourists. The increased investment in the neighborhood will lead to revitalization and gentrification. Such gentrification may also displace local businesses to make space for larger chain stores due to the increment of rent (Cohen, 2018). The rise in property prices brings in more affluent residents, thus, increasing the standards of living in the neighborhood. This could potentially bring about exclusion and segregation of different social groups. The original residents, especially those of low socioeconomic status, may be priced out in the process (Ding et al., 2016). The effects displace the social fabric and cultural practices of the neighborhood, altering the culture and character of the place. The original residents may no longer feel a sense of belonging and community identity.

4 Cultural Districts for Tourism—The Case of WKCD

4.1 WKCD's Initiatives in Boosting Cultural Tourism

To fully understand the potential of cultural districts for tourism, this section examines the strategies implemented by WKCD in propelling the district into a major cultural attraction/destination. As most of the current landmarks within WKCD were launched during times of adversities in Hong Kong, where inbound tourists' numbers were significantly low, it is no surprise that the district focused all its attention then in engaging residents and communities. In May 2021, in view of social distancing, WKCD rolled out several social media initiatives, which included animating the murals at Freespace and converting them into Instagram filters and WhatsApp stickers. Similar efforts were also put in for the Xiqu Centre. These initiatives catered well to the youths of Hong Kong in shaping perceptions of the cultural district, particularly Xiqu Centre, as its focus on traditional Chinese opera is deemed by youngsters as traditional and boring (Ho, 2017).

During the first year of M+'s opening in November 2021, the museum offered free admission. This strategy not only attracted large numbers of visitations amongst residents and communities, but also demonstrated WKCD's objectives to democratize culture. M+ introduced an admission fee on its first anniversary but maintained one exhibition, 'Hong Kong: Here and Beyond', free to the public until June 2023. This move was strategic in the sense that it encouraged public participation, but also ensured continued revenue stream to sustain the museum. Furthermore, keeping this exhibition free allowed more Hong Kong residents and tourists to learn about the city's unique visual culture, which includes art, design, film-making, and architecture. HKPM, on the other hand, did not extend the same offer when it first opened in July 2022. Nevertheless, the museum distributed over 200,000 sponsored tickets to students and underprivileged communities during the first month of operation. All in all, the efforts to engage residents and communities in times of adversities have proven to be fruitful. The initiatives fostered a strong sense of pride and ownership of the cultural district amongst the locals, and in turn, the locals have unconsciously become ambassadors in promoting WKCD to their local and overseas peers.

A major event by M+ that created much hype in Hong Kong was the launch of the special exhibition 'Yayoi Kusama: 1945 to Now' in November 2022. Kusama is a world-renowned Japanese contemporary artist, known for her colorful and distinctive artworks that often feature polka dots and mirrored rooms. It is worth noting that the exhibition opened during a transitional period, where some COVID-19 restrictions had eased up but were yet to be fully relaxed. Inbound travelers were subjected to the 0 + 3 measure, where they were prohibited from entering certain premises, including museums, and performing arts venues. Then, Betty Fung, CEO of WKCD Authority, appealed to the government to allow inbound visitors to participate in WKCD's events, contending that such activities did not involve the removal of masks and, therefore, posed minimal health risks (Yeo, 2022). Fung's move was evident that WKCD needed to expand visitors' numbers and profiles.

Proactive measures were introduced to respond to the easing of border restrictions between Hong Kong and Mainland China in early 2023. At around the same time, HKPM rolled out a new exhibition titled 'Radiance: Ancient Gold from the Hong Kong Palace Museum Collection and the Mengdiexuan Collection'. The exhibition, which is still running during the time of writing, presents an exquisite collection of ancient gold artifacts spanning over 3000 years, offering insights into Chinese culture, history, and craftsmanship. The Xiqu Centre, on the other hand, has dedicated a substantial amount of its programming to Cantonese opera and other forms of Chinese opera. In particular, the Tea House Theatre in the center offers unique experiences, which serve to preserve traditional art forms on one hand, and appeal to new audiences on the other. These efforts highlight Hong Kong's position as a East-meets-West hub, and at the same time, dispel Hong Kong's image as a cultural desert. To attract more Mainland Chinese tourists, WKCD introduced a free loop bus service between West Kowloon Station, an important port of entry for Chinese tourists, and the cultural district, as well as water taxi services on

Saturdays with several stops along Victoria Harbor, providing a scenic route for visitors coming to the district.

Apart from its unique programs, WKCD's physical environment plays a key role in its appeal as an attraction/destination. The Xiqu Centre, Freespace, M+, and HKPM are cultural landmarks in their own rights with striking architectural designs. The Art Park also provides families (including pets) and friends a space to chill and enjoy the breathtaking views of Hong Kong's skyline, the harbor, and the iconic buildings in the district. These buildings and landscape provide visitors the opportunities for captivating photos, which are often posted on social media platforms. Social media influencers from Hong Kong, Mainland China, and the region with large numbers of followers are also shaping travel trends. Many shared images and engaging posts about WKCD. Collectively, these social media posts drive up the visitation numbers of the district.

Given its status as a cultural hub and leader, WKCD is in the position to not only identify but also address the gaps that exist within Hong Kong's arts and cultural sector. One such gap is the limited engagement and collaborations between major institutions and smaller arts and cultural organizations. Currently, smaller organizations tend to work in silos or within a certain clique with limited opportunities for collaborations, especially with major institutions such as WKCD. These organizations might seem trivial in their individual contribution to the arts ecosystem of a city. Nevertheless, they collectively form a vibrant backbone to a city's cultural and creative industries (Lim, 2022). At the same time, they promote cultural diversity and resilience amongst practitioners through their works and operations. Collaborations with these organizations bring about richer and more authentic cultural offerings, which are appealing to both residents and international tourists.

Another gap that can be addressed by WKCD relates to the accessibility and relevance of the arts to the public. While the district has put in considerable efforts in improving local visitor numbers, there is still a sizable public that has no access to arts and cultural activities. WKCD can invest in further research and come up with viable strategies to cultivate a more inclusive environment for all. One way for WKCD to go about creating inclusivity is to leverage on the distinctive culture of Hong Kong, which is shaped by the lived experience of its people. There could be a better balance between regional and international programs, alongside cultural activities that focus on presenting stories around local communities and their histories. An interesting observation is that the performing arts institutions, such as Xiqu Centre and Freespace, are more proactive in their collaborations with local artists and organizations, as compared to M+ and HKPM. This disparity may be due to differences in the artistic mission/vision of the respective institutions. Furthermore, due to the lack of performance venues in Hong Kong, the Xiqu Centre and Freespace venues are in high demand for rental by local artists and organizations. Nevertheless, there are still many available physical spaces in both museums that could be utilized for collaborations with local artists and organizations to generate intriguing programs, further enhancing the district's appeal as a cultural attraction/destination.

When exploring WKCD as a viable cultural tourism hub, an interesting point to note is in the different approaches towards marketing museums and performing arts events. In the context of Hong Kong, museum exhibitions last longer, have larger capacity, and cater well to accidental and casual cultural tourists who may not have a fixed itinerary. On the contrary, performing arts events have shorter runs in Hong Kong. Given the limitation of venue capacity and fixed timings of performances, performing arts events do not cater as effectively to accidental and casual tourists. It is also likely that the tickets might have been sold out by the time these tourists hear about the events. For performing arts events in the cultural district to be included in the larger cultural tourism landscape, efforts should be targeted towards purposeful cultural tourists. To do so, the cultural district must put in place an effective system that allows seamless data sync across different ticketing partners and platforms.

4.2 Untapped Potential of WKCD

A major roadblock affecting WKCD's potential in becoming a tourism hub lies in the fragmentation between the various institutions within the district. A recent case-in-point, which was reported in the news, relates to the admission policies of HKPM and M+. Since July 2023, HKPM has suspended the free admission offer on Wednesdays, citing the ineffectiveness of the initiative. Conversely, M+, despite having higher visitor numbers, reduced its admission fees by 25% for 2 months at the same time (Li, 2023). This example, though minor, goes to show the lack of cohesive efforts in pricing strategies for both museums. As mentioned repeatedly, there is no doubt that each of the institutions have rolled out well-curated cultural programs in recent years. However, there seems to be a lack of synergies and collaborations between the institutions, and this presents a missed opportunity for the district to showcase cross-cultural experiences to visitors. At this stage, it seems as if each of the institutions are standalone cultural landmarks, and their individual significance outweighs that of the district itself. While there is evidence of collective efforts in marketing WKCD, these initiatives cannot truly come to fruition without first having a compelling and viable product or experience representing the district. In arts marketing, the 'product' strategy is of ultimate importance. Promoting collaborations between the various institutions would propel WKCD towards becoming a unified cultural hub. One potential collaboration, for instance, can be for a district-level arts and cultural festival that is curated collectively by all the artistic heads.

On the same note, WKCD also faces shortfalls in its hospitality offerings. Each cultural institution manages its own guest relations, which is nonetheless crucial in supporting visitors faced with problems. However, there is a lack of the presence of a district-wide guest relations team that could support, and at the same time, curate experiences for visitors to explore the expansive cultural district. For instance, there are currently no clear initiatives to drive visitors of HKPM to visit M+ (or vice versa) even though the two institutions are located just minutes apart on foot. Beyond guest relations, there is also the absence of convenient and friendlier priced retail and cafes/restaurants in the district. To avoid paying hefty prices for food, visitors might only visit the district before or after mealtimes. Resolving these hospitality issues can improve visitors' satisfaction, lead to higher visitor numbers, and increase time spent in the district.

To cater to more tourists, WKCD can also further tap on the incorporation of digital tools. At present, WKCD's website primarily serves as an informational site. While such information is vital, there lies opportunities to create more immersive digital experiences for potential visitors. For instance, the website could incorporate interactive elements such as 360-degree view of the district and its various institutions and landscape. The virtual tour could help visitors visualize a physical visit and at the same time, learn more about the district. Each of the institutions can also consider presenting online performances/exhibitions and digitized archives/collections. M+ has taken the lead in this respect, with purposefully curated online exhibitions and collections accessible 24/7. Such online presence extends the reach of the institution and diversifies the modes of presentations, enhancing the viability of the product strategy of the district.

The interactive digital initiatives, such as Instagram filters, introduced during the COVID-19 period should not merely be a stopgap measure. During the pandemic, digital technologies played a key role in disseminating culture (Jeannotte, 2021). Online exhibitions, virtual tours, and digital storytelling have allowed global audiences to experience Hong Kong's culture from the comfort of their homes. While the tourism sector is back to physical interactions, the adoption of technologies during the pandemic offers valuable lessons to reshape the future of cultural tourism in Hong Kong. WKCD could further enhance lessons learned during the pandemic in creating a more digitally integrated experience for its visitors. One possibility is for the website to include interactive elements where online visitors participate in some form of co-creation with opportunities for their works to be displayed virtually or even onsite. On a similar note, these features can also be included on the websites of the respective cultural institutions.

The core issue confronting WKCD's full realization of its potential is its incompleteness. Over the last decade, the district has faced multiple delays in meeting construction deadlines. At this point, it is only known that the Lyric Theatre Complex will open sometime in 2026 and that conglomerate Sun Hung Kai has won the bid for a major commercial property project onsite. The public is unaware of other concrete cultural projects and plans in place for the remaining undeveloped plot of land in the district. This incompleteness certainly has an impact on the district's potential to become a destination for cultural tourism.

4.3 Stakeholders' Roles in Cultural Tourism

To become a hub for cultural tourism, WKCD's efforts, alone, are not sufficient. Inputs from diverse stakeholders are necessary in creating a vibrant and sustainable development of cultural tourism in Hong Kong.

One of the major stakeholders in cultural tourism is the government. To a large extent, the success of developing cultural tourism in particular locales lie in the government's attitude towards its values, coupled with the effective implementation of relevant policies. In July 2022, the government of Hong Kong established the Culture, Sports, and Tourism Bureau, bringing arts, culture, and tourism together under one governmental body for the first time. The establishment of this bureau is very much linked to China's broader policies, including the 14th Five-Year plan, which emphasizes on the significance of the cultural industries and their potential in contributing to the economy. Furthermore, several infrastructural developments, such as the Hong Kong–Zhuhai–Macau Bridge and the high-speed rail in Hong Kong, open new possibilities for more holistic cultural tourism experiences, especially between the Greater Bay Area cities. Apart from expanding the cultural offerings, these enhanced connectivity enables cultural tourists to gain more knowledge and appreciation of the region's culture.

To kickstart the development of viable cultural tourism, the government plays a crucial role in bringing together the different stakeholders and educating them on the potential of cultural tourism. From there, stakeholders can understand their respective roles and expectations. Although the Hong Kong government has formed the bureau focusing on culture, sports, and tourism, there is still a lack of a clear and actionable blueprint for cultural tourism. Developing a blueprint involves a thorough research process to identify market conditions, stakeholders' perceptions, and the uniqueness of Hong Kong's culture. The research outcomes are key to defining a viable strategic and action plan in driving cultural tourism. One aspect that must be considered in the blueprint is the balance between economic contributions and cultural integrity. While economic impact is of great importance for any tourism activities, ethics behind preserving and presenting Hong Kong's unique culture and heritage must be upheld.

WKCD serves as a great case for more cultural districts to be developed and promoted in Hong Kong, especially within existing neighborhoods. One great example is Sham Shui Po. With its encapsulation of a rich everyday lives of Hong Kong people, coupled with its diverse gastronomic offerings, unique shopping experiences (such as second-hand goods market), different architectural structures and monuments, as well as stories from local communities, Sham Shui Po is an ideal location for a community-based cultural district. These attributes of Sham Shui Po are also known as territorial capital, which refers to the set of localized resources that account for the potential of the locale (Camagni, 2009). In the context of Hong Kong, having several unique cultural districts not only cater to differ types of cultural tourists, but also showcase the city's diverse cultures. Collectively, the different cultural districts make up Hong Kong's territorial capital.

Another important government-related tourism stakeholder in Hong Kong includes the Hong Kong Tourism Board (HKTB), which is not only instrumental in promoting Hong Kong (including WKCD) on a global scale, but also plays a key role in cultivating cultural tourism in the city. HKTB's support of WKCD is visible on its website, where the district is prominently featured. This includes listing the ongoing events in WKCD and rolling out special features of the district, such as

'5 West Kowloon Cultural District must-sees'. Beyond the website and social media posts, HKTB has also included WKCD in their recent 'Hello, Hong Kong' campaign, which is aimed at generating global interests in Hong Kong as a destination. The promotional video of this campaign features Hong Kong pop star Aaron Kwok's high energy dance with M+ as the backdrop. This blend of popular culture with M+'s iconic architecture is symbolic in a way that it characterizes WKCD (and Hong Kong) internationally as a destination where arts and cultural heritage meet the contemporary cityscape.

Another crucial yet often overlooked stakeholder is the education sector, particularly higher education. This sector is instrumental in preparing future practitioners and researchers in the field of cultural tourism, specifically in areas such as policy making, heritage conservation, creative place-making, arts and cultural management, monetization of creative outputs, sustainable tourism practices, amongst others. Education institutions in Hong Kong should be encouraged and incentivized to develop new programs and conduct research related to cultural tourism. This could be achieved through direct subvention from the government, partnerships between the cultural tourism industry and institutions, and/or through encouraging interdisciplinary collaborations between tourism and humanities departments in the institutions.

Other important stakeholders driving the tourism sector include corporations/businesses, hotels, and tour operators. Hong Kong's flagship carrier, Cathay Pacific, has featured WKCD multiple times in its in-flight magazines and websites. These features introduce WKCD to international travelers and promote the district as a must-visit destination. Several local hotels have also rolled out guest packages that incorporate visits to M+, and in so doing, bridge the gap between tourists and Hong Kong's arts and cultural scene. The growing prominence of online travel agencies in the last decade has opened new avenues and practices in promoting tourism (Ma et al., 2023). WKCD has established partnerships with various online travel agencies, selling admission tickets and collaborating on feature articles on the happenings in the cultural district. Furthermore, such portals usually encourage customer reviews, which allow future visitors of the district to gain insights into the experiences and recommendations of those who have visited prior.

5 Conclusion

The term 'Cultural Tourism' is made up of two distinct yet somewhat intertwined words—'culture' and 'tourism'. An important consideration here is which of these two words should be given more precedence, or should they be given equal weightage? The establishment of the Culture, Sports, and Tourism Bureau is an important first step. Nevertheless, the bureau could be more expeditious in rolling out clear blueprints and policies for cultural tourism. The first policy address by the newly appointed Chief Executive John Lee in October 2022 reinforced the government's resolve. The address reiterated Hong Kong's position as the East-meets-West Centre for International Cultural Exchange and laid out broad directions in enhancing cultural tourism in the city. There have also been debates in the legislative council meetings about the potential of cultural tourism.

At present, there are no statistics informing the visitor numbers of the cultural district in its entirety. By the first half of 2023, both M+ and HKPM have collectively welcomed around four million visitors. These numbers, while encouraging, should not be the basis for celebration. The reality is that WKCD still faces hefty financial challenges. The initial endowment of HK$21.6 billion is expected to be used up in less than 2 years, and at the same time, the district is still operating at a loss. This situation calls for immediate actions in developing a viable and self-sustaining model of cultural tourism in Hong Kong. In developing cultural tourism, policymakers and stakeholders should also be aware of potential risks such as over-tourism. Cultural attractions can lead to an influx of visitors, which may threaten heritage conservation, put a strain on the resources, and more undesirably, upset existing community lives. It is hoped that this chapter contributes to the discussions surrounding the potential of cultural districts in boosting tourism in Hong Kong through exploring WKCD's strategies and challenges. While the paper focuses on Hong Kong, the arguments presented can inform the planning and development of cultural districts for tourism in nearby regions.

References

Arnaboldi, M., & Spiller, N. (2011). Actor-network theory and stakeholder collaboration: The case of cultural districts. *Tourism Management, 32*(3), 641–654.

Barber, L. B. (2019). Heritage tours and trails on foot in Hong Kong: Towards a typology that crosses the tourist-local divide. *Journal of Heritage Tourism, 14*(4), 295–307.

Borrup, T. (2013). Artists & creativity in urban placemaking: Reflections on a downtown Minneapolis cultural district. *Journal of Urban Culture Research, 7*, 96–111.

Camagni, R. (2009). Territorial capital and regional development. In R. Capello & P. Nijkamp (Eds.), *Handbook of regional growth and development theories* (pp. 118–132). Edward Elgar.

Chhabra, D. (2022). *Resilience, authenticity and digital heritage tourism*. Routledge.

Chiu, S. T., & Huang, W. J. (2022). Heritage interpretation from supply and demand perspectives: The case of culinary heritage workshops in Hong Kong. *Journal of Heritage Tourism, 17*(5), 563–592.

Cohen, N. (2018). Feeding or starving gentrification: The role of food policy. *CUNY Urban Food Policy Institute, 5*, 1–10.

Ding, L., Hwang, J., & Divringi, E. (2016). Gentrification and residential mobility in Philadelphia. *Regional Science and Urban Economics, 61*, 38–51.

Domínguez-Quintero, A. M., González-Rodríguez, M. R., & Paddison, B. (2020). The mediating role of experience quality on authenticity and satisfaction in the context of cultural-heritage tourism. *Current Issues in Tourism, 23*(2), 248–260.

Duhme, L. (2012). *Cultural tourism: Case study Portugal*. Diplomica Verlag.

Ho, E. (2017). *How Chinese opera helped HK students bond with their families and heritage, and why they want to keep the tradition alive*. 28 Sept 2017. https://www.scmp.com/yp/discover/lifestyle/features/article/3066380/how-chinese-opera-helped-hk-students-bond-their

Ho, P. S. Y., & McKercher, B. (2004). Managing heritage resources as tourism products. *Asia Pacific Journal of Tourism Research, 9*(3), 255–266.

Hong Kong Tourism Board. (2023). *HKTB announces 2022 full-year visitor arrivals*. 16 Jan 2023. https://www.discoverhongkong.com/content/dam/dhk/intl/corporate/newsroom/press-release/hktb/2023/01-2022-Full-Year-Arrivals-E.pdf

Jeannotte, M. S. (2021). When the gigs are gone: Valuing arts, culture and media in the COVID-19 pandemic. *Social Sciences & Humanities Open, 3*(1), 100097.

Karvelyte, K. (2018). From arts desert to global cultural metropolis: Tte (re)branding of Shanghai and Hong Kong. In V. Durrer, T. Miller, & D. O'Brien (Eds.), *The Routledge handbook of global cultural policy* (pp. 247–264). Routledge.

Large, J. (2019). Conspicuously 'doing' charity: Exploring the relationship between doing good and doing harm in tourism. In T. Raymen & O. Smith (Eds.), *Deviant leisure: Criminological perspectives on leisure and harm* (pp. 324–345). Springer.

Li, A. (2023). Hong Kong's M+ art museum to cut ticket prices by 25 per cent over summer despite art hub facing financial constraints. 5 July 2023. https://www.scmp.com/news/hong-kong/hong-kong-economy/article/3226627/hong-kongs-m-art-museum-cut-ticket-prices-25-cent-over-summer-despite-art-hub-facing-financial

Lim, B. (2018). Representations of Hong Kong's West Kowloon Cultural District by local english newspapers. *Southeast Asia: A Multidisciplinary Journal, 18*, 44–54.

Lim, B. (2022). Beyond size and scale: Reflections on small performing arts organizations. *RICE Journal of Creative Entrepreneurship and Management, 3*(1), 55–65.

Ma, S., Li, H., Hu, M., Yang, H., & Gan, R. (2023). Tourism demand forecasting based on user-generated images on OTA platforms. *Current Issues in Tourism*. https://doi.org/10.1080/13683500.2023.2216882

Mackellar, J. (2014). *Event audiences and expectations*. Routledge.

Markusen, A., & Gadwa, A. (2010). Arts and culture in urban or regional planning: A review and research agenda. *Journal of Planning Education and Research, 29*(3), 379–391.

McKercher, B., & Du Cros, H. (2003). Testing a cultural tourism typology. *International Journal of Tourism Research, 5*(1), 45–58.

Mihelj, S., Leguina, A., & Downey, J. (2019). Culture is digital: Cultural participation, diversity and the digital divide. *New Media & Society, 21*(7), 1465–1485.

Ng, J. (2009). *Paradigm city: Space, culture, and capitalism in Hong Kong*. SUNY Press.

Nguyen, T. H. H., & Cheung, C. (2014). Tourists' perceptions of authenticity: The case of heritage experiences in Hong Kong. In V. T. King & P. Porananond (Eds.), *Rethinking Asian tourism: Culture, encounters and local response* (pp. 71–94). Cambridge Scholars Publishing.

Noonan, D. S. (2013). How US cultural districts reshape neighbourhoods. *Cultural Trends, 22*(3–4), 203–212.

Noonan, D. S., & Breznitz, S. M. (2013). Arts districts, universities, and the rise of media arts. In M. Rushton (Ed.), *Creative communities art works in economic development* (pp. 118–143). Brookings Institution Press.

Nuccio, M., & Ponzini, D. (2017). What does a cultural district actually do? Critically reappraising 15 years of cultural district policy in Italy. *European Urban and Regional Studies, 24*(4), 405–424.

Pine, B. J., & Gilmore, J. H. (1999). *The experience economy*. Harvard Business School Press.

Ponsignon, F., & Derbaix, M. (2020). The impact of interactive technologies on the social experience: An empirical study in a cultural tourism context. Tourism management. *Perspectives, 35*. 100723. https://doi.org/10.1016/j.tmp.2020.100723

Puig, F., & Marques, H. (2011). *Territory, specialization and globalization in European manufacturing*. Routledge.

Ramchander, P. (2007). Township tourism: Blessing or blight? The case of Soweto in South Africa. In G. Richards (Ed.), *Cultural tourism: Global and local perspectives* (pp. 39–67). Routledge.

Richards, G. (Ed.). (2001). *Cultural attractions and European tourism*. CABI publishing.

Richards, G. (2021). *Rethinking cultural tourism*. Edward Elgar Publishing.

Sacco, P. L., Tavano Blessi, G., & Nuccio, M. (2008). *Culture as an engine of local development processes: System-wide cultural districts*. Università Iuav di Venezia.

Santagata, W. (2002). Cultural districts, property rights and sustainable economic growth. *International Journal of Urban and Regional Research, 26*(1), 9–23.

Santagata, W. (2006). Cultural districts and their role in developed and developing countries. In V. A. Ginsburg & D. Throsby (Eds.), *Handbook of economics of art and culture* (Vol. 1, pp. 1001–1119). North-Holland.

Santagata, W. (2011). Cultural districts. In R. Towse (Ed.), *A handbook of cultural economics* (pp. 147–152). Edward Elgar.

Smith, M., MacLeod, N., & Robertson, M. H. (2010). *Key concepts in tourist studies*. Sage.

Tam, L. (2023). *Why Hong Kong needs a culture tsar and not just a 'mega' events committee—To champion our local identity to visitors*. 4 Feb 2023. https://www.scmp.com/lifestyle/arts-culture/article/3208963/why-hong-kong-needs-culture-tsar-and-not-just-mega-events-committee-champion-our-local-identity

To, W. M., & Lee, P. K. (2023). The COVID-19 impacts on air transport-a case study of Hong Kong. *International Journal of Shipping and Transport Logistics, 16*(3–4), 256–276.

Webb, D. (2017). Trends in the development and operation of performing arts centers. In P. D. Lambert & R. Williams (Eds.), *Performing arts center management* (pp. 57–74). Routledge.

Yeo, R. (2022). *Hong Kong cultural district head appeals for tourists to be allowed to visit museums while under Covid 3-day medical surveillance*. 1 Nov 2022. https://www.scmp.com/news/hong-kong/hong-kong-economy/article/3197925/hong-kong-cultural-district-head-appeals-tourists-be-allowed-visit-museums-while-under-covid-3-day

Benny Lim is Associate Professor of Practice in Cultural Management and Director of the Master of Arts in Cultural Management programme with the Chinese University of Hong Kong. He is also a Visiting Professor with the Design School of Taylor's University (Malaysia). Prior to his current appointments, Benny was in the senior leadership team of Berjaya University College and KDU University College in Malaysia. He was a Visiting Scholar at the Peking University's Institute of Cultural Industries (2019–2020) and Honorary Research Fellow of University of Nottingham's (Malaysia) Asia Research Institute (2019). Currently, he is the Board Chair of Along the Edge Limited, a non-profit arts organization in Hong Kong, as well as a member the Programme Sub-committee of The TENG Company, Singapore. In January 2023, he was appointed as an Arts Advisor (Arts Administration) for the Hong Kong Arts Development Council.

Chapter 9
Red Tourism in Rural Beijing: The Hierarchical Governance and Grassroots Community Engagement

Ran Liu

1 Introduction

Red tourism is an example of the "politics of heritage" in tourism in China (Timothy, 2011; Zhao & Timothy, 2015). Using tourism as a tool can teach about cultures and histories, and at the same time instill political ideologies by the ruling class through propaganda and promoting people's devotion to the leaders and patriotism (see Timothy, 2007; Henderson, 2007). Red relics and heritage can be heavily politicized through tourism, owing to their potential to harbour social values, norms, and beliefs. Red tourism in China consists of mostly state-promoted tours to the revolutionary memorial sites that are related to the history of the Chinese Communist Party (CCP), legitimizing the party leadership and reinforcing the Chinese national identity. Red tourism now represents a prime tourism product in the Chinese tourism market. From 2004 to 2019, red tourism drew an increasing record of domestic tourists from 140 million to 1.41 billion (CN.Chinadaily, 2021). Even during the COVID-19 lockdown year in 2020, domestic tourists of red tourism still reached one billion (Wang, 2023). During 2021 at the commemoration of the CCP's 100-year anniversary, the state-sponsored celebration activities nationwide further spurred the popularity of red tourism. It was found that participants from the "Generation *Z*" (born from 1995 through 2010) accounted for 44.56% of the red tourists during the summer of 2019 (*ibid*).

Indeed, red tourism planning and marketing in China is driven by the central government. As early as 2003, China National Tourism Administration (CNTA) began to promote "red tourism" as a nationwide education campaign to glorify the

R. Liu (✉)
College of Resource Environment and Tourism, Capital Normal University, Beijing, China

T.-C. Wong et al. (eds.), *Cultural Tourism in the Asia Pacific*,
https://doi.org/10.1007/978-3-031-63459-8_9

CCP's revolutionary history. Then in 2004 the central government,[1] under the command of the National Red Tourism Coordination Executive Team (NRTCET), launched the "National Red Tourism Development Plan (2004–2010)" as a stimulant for market-led growth where "red tourism" was defined as a themed activity involving history learning activities and sightseeing at revolutionary heritage sites. NRTCET delineated different regions across the country with specific themes for the selected sites and funded infrastructure development of places where physical access was difficult. In publicity, NRTCET published a series of books such as the "Annual Reports on the Chinese Red Tourism Development" from 2005 to 2015 and other pamphlets (National Red Tourism Coordination Executive Team, 2005–2015).

Under the strong state propaganda campaign, a large number of memorial buildings and study materials have been established. From the mid-2010s till now, famous red tourism attractions, such as Mount Jinggang, have been heavily researched in the areas of revolutionary paradigm, cultural and combatory aspects of the Red Army's struggles in the past, among others (see Gu et al., 2007; Zuo et al., 2017; Tang et al., 2021). Hung (2018) in his publication introduced in particular the Five Sacred Revolutionary Sites with lots of details, namely Shanghai's First National Congress of the CCP, Mount Jinggang in Jiangxi Province, Zunyi in Guizhou Province, Yan'an in Shaanxi Province, Xibaibo in Hebei Province. Hung's work touched upon the key heritage features of red tourism. These include the historical museums and memorial halls, key battlefields and other known historical sites, former residence of Mao Zedong and other leaders, and the memorial parks in remembrance of the revolutionary martyrs. Politically, these heritage sites of red culture are designated to serve as the basis of educational patriotism aimed at the young generations.

Red tourism is a top-down ideological infusion by the CCP into heritage tourism, which is largely rural-based as the sites are spread out widely in remote rural China. Its target has recently been incorporated at different government levels into the Rural Revitalization and Rural Poverty Alleviation programs to help alleviate incidence of poverty via a top-down political hierarchical organizational network (see Fei, 1992/1947; Chen & Chen, 2004; Chen et al., 2013). Taking Gutian in Fujian Province as a case study, Zhao and Timothy (2015) discussed the major roles played by the local government in promoting red tourism at the meso- and micro-levels using both top-down and bottom-up approaches.

More recently, Wang et al. (2022) also investigated three typical red tourism destinations—Zunyi and Chishui in Guizhou Province and Panzhihua in Sichuan Province. In their study, they revealed two streams of conflicts in the management style in these red tourism destinations. These two conflicts were characterized by the difference in enforcement rigour between the central government and the local government, and views between local government officials and other stakeholders involved in the red tourism business.

[1] The General Affairs Office of the Central Committee of the Chinese Communist Party (CCP) and the General Office of the State Council of China.

But few studies were undertaken in unknown villages. This study examines how unknown villages with revolutionary records could be made to become a shining star in red tourism development. Using the empirical evidence from Taopeng Village in Pinggu District of Beijing, this chapter analyses how the grassroots Village Committee has mobilized the party organs to rebrand a deep-mountain poorly equipped village with less than 100 rural households, and reported success. Taopeng Village's experience and governance practice thus deserve further investigation.

2 Red Tourism and Its Nature

2.1 Red Tourism as an Ideological and Politicized Function

The ideological content and political input of red tourism in China deserves much debate (Gao & Guo, 2017; Li & Hu, 2008). As argued by Li and Hu (2008), the way red tourism is being practiced in China is strongly politicized when they compared it with that of the former communist countries in Eastern Europe such as the former East Germany, Romania, and Hungary. Firstly, the Chinese approach is driven by the Chinese Communist government aimed at its own nationals whereas the former Eastern European Communist states targeted their clientele at foreign visitors, notably those from the West (Light, 2001).

Secondly, red tourism in China is a political strategy to stimulate a collective nostalgia within its own people of the movements of Chinese communism in the past, with an objective to justify the legitimacy of the Communist rule. But as discussed by Coles (2003), communism-related heritage tourism in Eastern Europe was intended to reconstruct a collective identity of Europeanness. In contrast, red tourism in China could be defined as a politicized form of heritage tourism—laden with an ideological essence of red spirit or red memory, which is a combination of communist ideology, traditional Chinese virtues (such as collectivism), patriotism and national identity (Li et al., 2010).

Thus, the red tourism development in China may be interpreted as a continual process of cultural production and ideological evolution, as Gao and Guo (2017) probed red tourism into the politics of consumption. Althusser (1971) and Williams' (1977) notion on "ideology" is cited in Gao and Guo's study where they enquired whether a consumption practice in red tourism can be interpreted as a political act. As perceived by Althusser (1971), state apparatus can have two types: repressive and ideological. The "ideological" state apparatus can stretch into the public and the private domains, and can have a plurality of forms in publicity using radio, television, newspaper, literature, arts, sports, and entertainment including tourism (see Althusser, 2014; Gao & Guo, 2017). Consequently, tourism culture and tourist activities are material tools and practices that could be infused with an "ideology". Tourism in daily life can function as a platform, wherein the ruling party can propagate its ideology through which power could be maintained.

Besides, Williams (1977), in interpreting Marxist theory on "culture" and "ideology", his argument is that culture itself encompasses political, economic, and social activities. In this sense, red culture can be viewed as a material constitutive of a social process, through which the ideological relations are reproduced and experienced (ibid). Hence, the red cultural production, red memory construction, and ideological evolution are continual processes, which are governed by the leadership of the CCP with an implicit intention to inculcate ideological values into the tourists' personal experience. Althusser and Williams have thus reminded us that a better understanding of the behaviour of government agencies which influences consumer practices should be located within the larger political-ideological structure and context.

The "ideological" essence of red tourism in China is also reflected in the institutional trust from the local community support for the "red tourism" destination development. Zuo et al. (2017) used the data from residents living near Mount Jinggang to examine the effects of residents' institutional trust, especially the trust in the central government. As elaborated by Zuo et al. (2017), this community support is a kind of "substantive rationality" (as one form of the Weberian notions of rationality) which is common in hierarchical societies (Weber, 1978). "Substantive rationality" (instead of economic rationality) leads to a greater acceptance of authoritarian decision-making. That is to say, the residents' support may depend more on faith and affection for the government in a specific ideological-cultural structure. This influence of authoritarian values on the local community support for the "red tourism" development is especially prominent, since "red tourism" is a nationwide political campaign in China (Zuo et al., 2016).

2.2 *Red Tourism as a Means of Rural Poverty Alleviation*

As said, red tourism has a strong ideological input, but it is also integrated with a pragmatic and economic practice aimed at alleviating incidence of rural poverty. As early as October 1927, Chairman Mao established the Chinese Soviet revolutionary base in the remote Jinggangshan rural area where the Chinese revolution was characterized by rural and guerilla-based warfare which placed emphasis on encircling the city and seized power by force. To this end, some very remote mountain villages with a very small population have become *par excellence* red tourism destinations. As these destinations have hitherto remained poor rural areas, they are comfortably incorporated into China's prevailing "Rural Revitalization and Rural Poverty Alleviation programs". Such political initiatives include not only the red culture and red tourism that can play an agency role to strengthen the ruling Party's mandate associated with the national identity and party identity but also the rural economic growth in the former revolutionary bases. These sites are often isolated in the deep mountains with a small poverty-stricken population and poorly access infrastructure.

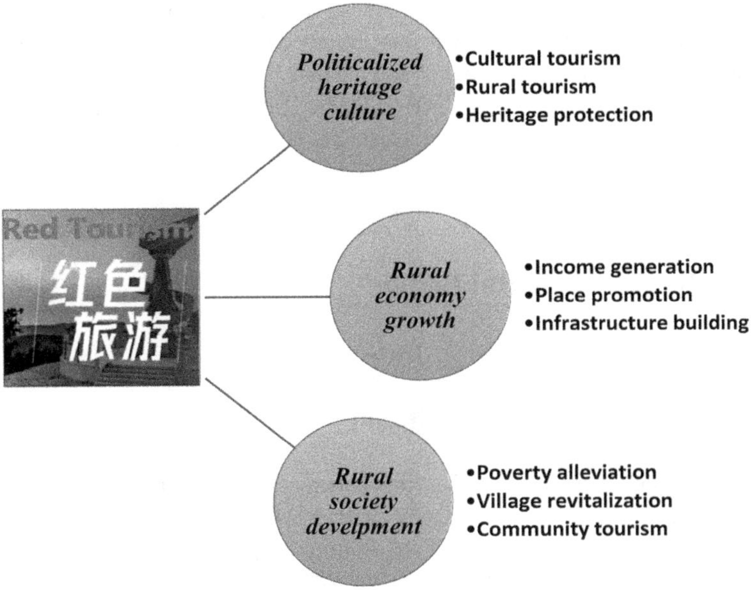

Fig. 9.1 The triangular relationship between politicized heritage culture, economic development and community building

Hung's (2018) study pointed out there is a development paradox between ideological motives and economic motives in the red tourism industry. The local community cadres understand that tourism investors have a preference for nonpolitical entertainment programs to those heavily infused with ideological agenda. For business investors, red tourism is not a merely political undertaking; it is also a place where tourists can relax and have fun. As a compromise, local grassroots government officials have come to terms to employ the tactic to promote Red-cum-Green Tourism. For instance, Huairou District in the northern Beijing rural outskirt had designed a one-day-tour allowing tourists to visit both revolutionary sites and some scenic points such as the Great Wall, a temple, and a forest park. Red relics are at the top of the must-visit list, and the green components are also highlighted (Beijing.gov.cn, 2021).

In the 2017 World Tourism Organization (UNWTO) Forum, five key areas were recommended to achieve sustainable development: (a) inclusive and sustainable economic growth; (b) the social inclusiveness, employment, and poverty reduction; (c) resource efficiency, environmental protection and climate change; (d) cultural values, diversity and heritage; and (e) the mutual understanding, peace and security (see Pan et al., 2018; UNWTO, 2018).

Roughly based on the above UN's notion, this study has identified and adjusted three key elements in the Chinese context in response to China's red tourism aimed at contributing towards a transformative change through rural tourism development. Figure 9.1 gives a conceptual relationship between politicized heritage culture, economic development and community building in the rural-based red tourism.

This Chapter uses a case study of Taopeng Village in rural Beijing to look into how the grassroots village committee has promoted and safeguarded their "red heritage" and green elements through lobbying and marketing strategies. During a "patriotic education" trip, a survey was conducted there with the participation of the human geography undergraduate students. The next Section will elaborate on the survey methods and the main findings.

3 Red Tourism in Rural Beijing

3.1 Survey Methods

3.1.1 Patriotic Education Trip in Beijing

Beijing is the capital city rich in red symbolism and history. In the celebration of the 100th anniversary of the CCP's foundation in 1921, the Publicity Department and the Party History Research Office of the Beijing Municipal Party Committee published jointly *The Map of Red Tourism in Beijing* with the Municipal Bureau of Culture and Tourism in Beijing. This book lists and maps out a total of 237 red tourism spots in Beijing, including the Patriotic Educational Bases, Cultural Relics Protection Units, and other attractions. Our fieldwork in this study chose a different site outside their list.

As part of the human geography field trip for an undergraduate class, we selected a red tourism site to integrate the course objective with ideas of patriotic and ideologically politicized education in response to the programme requirement (see Nairn, 2005; Hope, 2009). The human geography fieldwork gives students direct access to the "real world" where they could witness preset examples to enhance their level of understanding about China's revolutionary history. Using this patriotic education project to enrich the undergraduate students' experience in rural Beijing, Taopeng Village is an ideal site for such purpose.

There is a historical account which supports the choice of Taopeng Village. In 1939, the Communist Party leaders of Ji-Re-Cha Region Committee executed a "Trinity Campaign" in the Anti-Japanese War to the north of Beijing City. In this campaign, three main Anti-Japanese Bases were established around Beijing city, namely Jidong, Pingxi, and Pingbei (see Sites 1, 2, and 3 in the left map of Fig. 9.2).

3.1.2 Strategic Importance of Taopeng

It was in Taopeng Village of Pinggu District where the Communist First Party Branch Office was set up as a command centre in the Anti-Japanese campaign (see Site 1 in the left map in Fig. 9.2). It was here in this village where we as lecturers organized the students to conduct household surveys. Through this fieldwork, both lecturers and students had all deepened their understanding level of the past experience of this red tourism village. There were positive impacts on the students' cognitive acquisition and "direct experience" in this field trip. They have now a better understanding of the hardship the poorly equipped Communist fighters went through in the remote mountain Taopeng Village with less than 100 households (see land-use of Taopeng in Fig. 9.2, map on the right).

Another two First Party Branch Offices were equally established nearby. One was in Beijing's western suburb in Tianzhuang Village, Mentougou District, which was also known for being an important base linked to the Pingxi Revolutionary event. Our team visited this red heritage in the spring of 2023 (see Fig. 9.2, Site 2, map on the left). The other was found in the north of Beijing in Miaoshang Village in Huairou District. We visited with the students the Piposhi Village in Dazhuangke Town in Yanqing District, Beijing (see Fig. 9.2, Site 3, map in the left).

From our site visits, we have chosen Taopeng Village as an exemplary case of red culture and red tourism in this study. This is because Taopeng was branded a shining star by the Beijing Municipality in its red tourism development. Marvell and Simm (2018) suggest that Beijing as a global city, its choice should carry a significant implication in re-contextualizing Taopeng Village within a wider set of ideas. We feel it is worthy to investigate further as to how such a small village had been miraculously selected.

3.2 *Taopeng Village: The Case of a Shining Star*

3.2.1 Lobbying the City- and District-Level Leaders

The Chinese revolutionary history was characterized by guerilla-based warfare based on the rural revolutionary bases in the mountainous areas. Beijing is no exception. Figure 9.2's map on the left demonstrates the well-preserved sites and relics of the "First Party Branch" in rural Beijing (Taopeng Village in Pinggu District, Miaoshang Village in Huairou District, and Tianzhuang Village in Mentougou District). With a rich collection of relics including the annals of Beijing's revolutionary history on display, they are in a good position to provide tourism elements in this rural and green setting to attract visitors encouraged by state-led publicity with a patriotic history.

Starting in 2004, red tourism was launched by the CCP officially as an ideological-political project to consolidate and justify the legitimacy of the party rule by promoting patriotism nationwide and implementing it via high ranking officials' top-down manner. These officials have acted as planners and designers, resource organizers and providers, and coordinators nationwide.

Under the present Chinese system of hierarchical governance, local grassroots development opportunities are largely dependent upon political decisions of the higher level authorities. In this red tourism case, for instance, a bottom-up lobbying effort must be made to gain the support of the higher authority (see Fig. 9.3, Photos **a** and **b**). Consequently, the grassroots village leader must demonstrate his/her

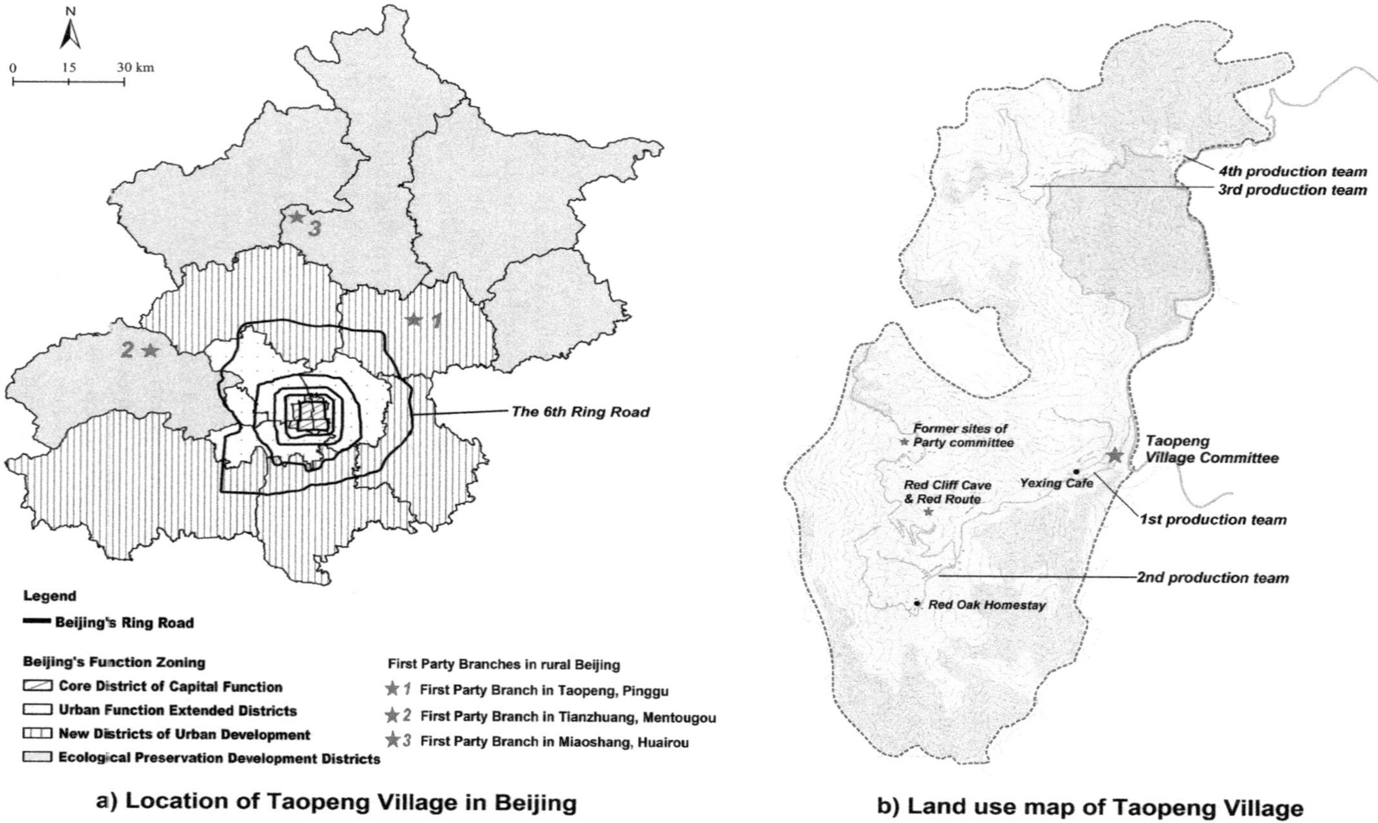

Fig. 9.2 Taopeng Village's location (left) and Land use (right). (Source: Land use map is from Taopeng Village Committee in 2021)

strong capabilities in running the campaign and in lobbying his/her senior officials and winning their hearts and minds (see Zhao & Timothy, 2015; Luo & Bao, 2019). In other words, the local grassroots leaders must act as creative entrepreneurs when they exercise their bottom-up organizational network, acting not only as a product manager, implementor, but also a relationship (*guanxi)* builder.

Taopeng was once the Revolutionary Headquarters during the 1930s–1940s. In order to attract tourists, red historic remains have been rehabilitated or rebuilt such as the First Party Branch Office housing the Jidong District Committee, Jidong Joint County Committee, and Public Security Department (see Fig. 9.3, Photo **c**). The sculptures of revolutionary leaders have been built in the "Red Cliff Cave" in commemoration of the pioneer leaders who worked there in 1940 in Pinggu, Beijing (see Fig. 9.3, Photo **d**). A tall monument was also built as the martyr cemetery, together with a themed revolutionary museum built nearby (see Fig. 9.3, Photos **e** and **f**).

The museum displays photos and maps of the revolutionary routes of struggle and acts of patriotism (see in Fig. 9.3, Photos **f** and **h**). Visitors to the museum can share the war-time experience with guided tours, storytelling, wearing Red Army suits and shoes. They can even taste the Red Army meals and sing patriotic songs, thereby sharing a personal experience lived through by the founding leaders (see Fig. 9.3, Photo **g**).

3.2.2 Mobilizing Group Tourism Demands from Beijing Municipality

Having been picked as a prototype for red tourism, organizing group tourism with large numbers of visitors is what needs to be followed up. In response, Taopeng Village has employed various strategies to attract visitors, including the mobilization of *danwei*[2] and public and private corporations, and schools to organize trips. Communist Party committees in *danwei* and companies are mobilized whereby political and patriotic education activities are organized. These periodically organized activities aim to bolster people's morale towards the love of the fatherland and the state (see Yan & Huang, 2017; Han, 2015).

To achieve the objective, the CCP organizing agencies set up an institutionalized system for financing their organizational activities (Koss, 2021). In recent years, latest means of technology is also employed such as the use of artificial intelligence and digitalization towards Party consolidation (*zhihui dangjian)*, internet-based Party building in the digital cloud (*dangjianyun*). Many of the Party's digital apps

[2] *Danwei* was the dominant institution and urban workplace in the public-oriented and distributive economy. *Danwei* acted as the state agencies in cities, responsible for the subsistence, political life, and social welfare (e.g., housing, medical care and kindergarten) for its employees. There were mainly three kinds of *danwei* in China: (a) *jiguan danwei* consisting of Party organizations and the governmental functional departments that hold the administrative power; (b) *shiye danwei* including non-profit units without administrative, allocative or regulatory authorities; and (c) State-Owned Enterprises (SOEs).

a) Beijing Mayor's calligraphic inscription

b) Still hanging in Taopeng's village office

c) First Party Branch site in Pinggu

d) Red Cliff Cave with the sculptures inside

e) The monument and martyr cemetery

f) The themed revolutionary museum

g) The trail of Anti-Japanese War

h) Anti-Japanese War Museum at Yuzishan

Fig. 9.3 The revolutionary relics in Taopeng (Photos taken by the author in 2021). (**a**) Beijing Mayor's calligraphic inscription (**b**) Still hanging in Taopeng's village office (**c**) First party branch site at Pinggu (**d**) Red Cliff Cave with the sculptures inside (**e**) The monument and martyr cemetery (**f**) The themed revolutionary museum (**g**) The trail of Anti-Japanese War (**h**) Anti-Japanese War Museum at Yuzishan. Note: Photo (**a**) is derived from the website: https://www.sohu.com/a/418154328_120209831

such as Weibo, WeChat, and other online platforms are actively deployed to strengthen the Party organs' openness and agility.

The deployment of net mechanism fits in well with the daily experience of the young Net Generation who are largely computer literate. As "digital natives", these Net Generation young people could easily be led to immerse themselves in digital technology all day long, imbuing them with sophisticated technical skills and learning preferences (see Bennett et al., 2008; Dingli & Seychell, 2015). There is however a significant difference between young people and older generation in technology skills, knowledge, and interests. For the young Net Generation, they are not satisfied with a static display of the museum objects. As such, participative exercises have been developed to provide the young with a one-day immersive experience in Party building activities which are posted on the Chinese social network Sina Weibo (Weibo, 2022).

Moreover, other participative activities are also organized. One is the wearing of Red Army suits and shoes, walking through the hilly terrain to the Red Cliff Cave (the birthplace of Pinggu's First Party Branch), and taking the oath when joining the Communist party as a member. The other is the walking of a 700-metre tour along the Anti-Japanese War trail frequented by the Party and military leaders, Red Army soldiers in the past, among others.

> The year 2021 is the CCP's 100-year anniversary. Many *danwei*s and companies are unable to leave Beijing during the epidemic period. Taopeng becomes a good site for the Party building. A dozen of *danweis* come to visit Taopeng each week. The tourist number of *danwei* or company visits and school trips is about 40 to 50 people each time. There are also Party school training classes with a large team of 100 to 200 people and some small groups with only five or six people. Some festivals with special commemorative significance, such as the founding of the Party and the Army Day, will attract more people. To ensure the high visibility of Taopeng, the village committee has combined old with the new media, such as the WeChat official site to attract the *danwei*s or schools. This further expands our influence through these "repeat visitors". (Investigation on Taopeng Village Committee, July 2021)

3.3 *The Economic Sustainability Aspect of Red Tourism*

3.3.1 Red-Green Tourism in Taopeng as a Sustainable Tourism Issue

As suggested earlier, rural regeneration is another key objective in the development of "red tourism" in the deep mountain areas in rural Beijing. The "First Party Branch Offices" in the villages like Taopeng are among the poorest and most remote in location. Their household incomes are among the lowest in the Beijing Municipality, and many youngsters have left for urban areas. As shown in Fig. 9.2, Pinggu District is located in the northeast of Beijing and Taopeng is a typical hilly village with almost 98% of it covered by forest (see Fig. 9.2). In 2021, Pinggu District had a few villages averaging 421 households with 1184 villagers (Pinggu Statistical Bureau, 2022). Among them, Taopeng Village had a much smaller population than other villages, and its population has shown a falling trend in recent years (Table 9.1). In 2021, 71 households with 131 people lived in Taopeng. Most of its residents were 70–80 years old. The working adults were predominantly those in their fifties and seven or eight households in the village owned fruit trees in the mountainous area, and most of the arable plots were not cultivated with crops.

Rural tourism in Huairou, Pinggu, and Mentougou Districts is the combination of their natural beauty (green) and the revolutionary heritage sites (red, see Fig. 9.2). Like many mountainous tourist sites in the north of Beijing, their aesthetic values are all well recognized in terms of clear air and scenic surroundings. Tourist experience here is supplemented by "green" views and mountain walks and there is a self-driving corridor leading to sites for mountain climbing. Figure 9.1 shows a typical route which connects the red tourism villages to green tourism sites. This is how the concept of Red-Green Tourism is derived, and this green economy has been used to serve the objective of rural poverty eradication in China in response to the call to achieve the UN Millennium Development Goals.

3.3.2 Potential Tensions: Collectivism in Red tourism vs. Individualism in Consumption

In their fight against local poverty, local party officials and the Taopeng village community have played an important role to transform their deep mountain settlement into a branded red village. One important aspect of economic development is their fast growing tourist accommodation facilities. According to our 2021 survey, Taopeng had comfortable and chain hotels offering beds and meals (see Fig. 9.2, the Red Oak Homestay). Growth in tourism is attributable to the quality provision of hotel accommodation and catering services as well as branding efforts, improvement in access roads which have attracted the incoming of middle class tourists from Beijing. Our field study shows some bed-and-breakfast inns in Taopeng are of high professional standard. With rising business opportunities, higher-end resorts and café are under construction in Taopeng (Fig. 9.4).

In the red tourism business, investors consist of a combination of local, extra-local and those initiated by the local Party Branch Office with the support from the Beijing municipal government. For local peasants, their focus is primarily the "Delights of Farm Guesthouse" (*nongjiale).* To the private investors from outside, their interest is focused on the higher-end resorts and café; Red Oak Homestay and

Table 9.1 The population and agricultural outputs in Taopeng Village (2018–2021)

Taopeng village	No. of rural households	Rural population (persons)	Agricultural employment (persons)	Total grain production (tons)	Fruit production (tons)
2018	118	217	115	0	279
2019	116	218	115		0
2020	112	217	115	0	22
2021	71	131	65		8
The average level in Pinggu's villages					
2021	421	1184	707	155	727

Source: Pinggu Statistical Bureau (2019–2022)

Yexing Café are typically their investments (Figs. 9.2 and 9.4). They have advertized their investments on social media pages, such as Tik Tok, WeChat, Weibo, and Xiaohongshu. For instance, Yexing Café is recommended by the Red Book (Xiaohongshu) in the Chinese social media and e-commerce platform (Liu, 2021). When potential young tourists browse visual images on social media (see Fig. 9.4, Photo **f**), they are likely attracted by the iconic and attractive photos, giving them the inspiration for a visit to such "perfect" locations by imagination (Liu, 2022; Guo, 2022). In this sense, the higher-end resorts and café are intertwined with a "consumer culture" that is circulated in social media spaces in China. Young people,

a) Red Oak Homestay and courtyard styles
b) Peasants unwilling to lease out their plots
c) Grape trellis in a new homestay courtyard
d) Cloud House Homestay under construction
e) Yexing Café as an internet-famous site
f) Yexing Café recommended in *Xiaohongshu*

Fig. 9.4 The higher-end resorts and café. (Photos taken by the author in 2021)
(a) Red Oak Homestay and courtyard styles (b) Peasants unwilling to lease out their plots (c) Grape trellis in a new homestay courtyard (d) Cloud House Homestay under construction (e) Yexing café as an internet-famous site (f) Yexing café recommended in *Xiaohongshu*

especially college students, are performing their ideal and beautiful "self", taste, and lifestyle—the consumers (as bloggers and tourists) advocate "individualism" that is distinct from the collectivism-oriented political consciousness in the red tourism (see Zuo et al., 2016).

In summary, rural tourism consumption in general represents a form of "consumerism" and "individualism" that is beyond the State's plan on the public-sector, including red tourism. As demonstrated in our in-depth investigation, Taopeng Village focuses on the planning and promotion of red tourism from the industry and region promotion perspective. For instance, Taopeng has attracted private investments. The private enterprise sector is continually sprawling in "First Party Branch" villages like Taopeng, but out of the State's interventionist policies in red tourism geared towards improving the revolutionary history of struggle and patriotism.

While red tourism is a strongly State-sponsored one-dimensional tourism product, Taopeng is equipped with a combination of local scenic values and higher-end resorts to satisfy middle class consumption demand (such as the boutique homestay and leisure café). Such packaging and integration of multi-dimensional products would serve better the changing generations (such as Net Generation and bloggers) in an increasingly open world. This transition is characterized by the passing of the older revolutionary generation with an ideological sentiment "robbing the wealthy to help the poor" to the present form of urban middle class deploying their consumptive power to support the rural poor.

Red tourism may still perform an important mythic role in addressing social harmony where the market economy permits entrepreneurial activities and wealth generation (Gu et al., 2007). Different economic sectors in this tourism business have played different roles in household income generation. At present, each household can earn "tourism dividends" (*fenhong*) from the village collectively run leisure business. Another source of collective revenue comes from the public subsidized ecological reservation policy, which compensates local farmers for keeping forest from being exploited for commercial use. In 2019, the subsidy set by the local government was 1350 yuan per *mu* (1 *mu* = 1/15 hectare) annually, with an annual increase of 50 yuan per mu. We were told by a village cadre that a household could earn 20,000–30,000 yuan per year. Many local peasants aged 50–60 have also been re-hired to manage the preserved forest to increase their income. The village committee told us also tent camping business can be set up in the forest under conservation to give tourists an option to spend their nights in the open air, and this is known as the "under-forest economy" (Linxia jingji). All in all, we were told that red tourism revenues made up about 70% of local household incomes. The rest of the income came mainly from running homestays and keeping bees to collect honey (Fig. 9.5).

Fig. 9.5 The livelihood of local peasants. (**a**) Local peasants' *nongjiale* (**b**) The peasants' vegetable picking orchards (**c**) Local peasants' beekeeping (**d**) Working in nearby vegetable greenhouses. (Photos taken by author in 2021)

4 Conclusion

Red tourism in China appears to be a political strategy to stimulate patriotic education and justify the legitimacy of communist rule (Gu et al., 2007; Timothy, 2011; Zuo et al., 2016). Previous case studies on the famous red tourism attractions, such as Mount Jinggang, as well as those highlighting the goals of red tourism were realized by the multi-tier government departments (Zhao & Timothy, 2015). This Chapter has addressed how unknown villages could become a shining star in the course of promoting red tourism development. Our fieldwork in this study chose a site outside the officially set must-visit list with an objective to showcase how human geography students could be taught patriotic education via field trips which have been integrated with the state-led education program (see Perrin & Gillis, 2019). In 2021 and 2023, two undergraduate human geography field trips were conducted respectively in three main Anti-Japanese Base Areas in rural Beijing, covering fully Pingxi, Jidong, and Pingbei rural sits around the Beijing city (Left map in Fig. 9.2). Taopeng is a remote mountain village rich in historical symbolism of revolutionary relics of the "First Party Branch" and the guerilla-based army in Pinggu (Right map in Fig. 9.2). The in-depth study is undertaken as to how a deep-mountain poorly equipped Taopeng village with less than 100 rural households could have been miraculously branded a shining star in the red tourism development. From our study, we have understood that where the village committee had successfully lobbied the city- and district-level leaders, and at the same time mobilized its own committee members to meet the red tourism demands with the help of public media and Beijing's Party quarters' networking, such achievement could be made.

Our study shows the red tourism in Taopeng has followed closely the norms set out by the framework of Sustainable Tourism for Development of the World Tourism Organization (UNWTO), as we can see in Fig. 9.1. Consequently, Taopeng has been assessed as an exemplary case of patriotic education in Beijing, China. With Taopeng rebuilt into an open museum and a lively classroom as an ideological apparatus to promote the education of patriotic and revolutionary traditions, red tourism has beefed up household income and revitalized the village community in a new mode of tourism activity.

However, there is a potential conflict between the "collectivist tourism" advocated in red tourism and the "individualistic business interest" inherent in this red tourism. On the one hand, there are the sprawling of higher-end resorts and café, characterized by higher-end consumption intertwined with a "consumer culture". On the other hand, with a great differential in individual entrepreneurship and business networking, the income gaps between individuals are inevitable. As young middle classes with high-end taste and consumption lifestyle taking place in the village, income and consumption pattern are somewhat distinct from the collectivism-oriented political consciousness in red tourism and the market-oriented pattern. The new challenge thus lies with how well the grassroots village committee can reconcile such a conflict between large external investors and other local stakeholders.

Acknowledgement This research was funded by a grant from the National Natural Science Foundation of China (NSFC Project number: 42071201, 41701188), and another grant from the Beijing Philosophy and Social Science Planning Program (19FXB024) and Social Science Research Key Program of Beijing Municipal Commission of Education (Project number: SZ202010028016). I gratefully acknowledge the cooperation from Professor Wang Pengfei for the investigations in 2021 and 2023 in rural Beijing.

References

Althusser, L. (1971). *Lenin and philosophy and other essays*. Monthly Review Press.

Althusser, L. (2014). *On the reproduction of capitalism*. Verso.

Beijing.gov.cn. (2021). *The one-day-tour recommendation of red tourism in Huairou District*. 20 April 2021. http://www.bjhr.gov.cn/ywdt/mtgz/202104/t20210430_2379077.html

Bennett, S., Maton, K., & Kervin, L. (2008). The "digital natives" debate: A critical review of the evidence. *British Journal of Educational Technology, 39*, 775–786.

Chen, X. P., & Chen, C. C. (2004). On the intricacies of the Chinese Guanxi: A process model of Guanxi development. *Asia Pacific Journal of Management, 21*, 305–324.

Chen, C., Chen, X., & Huang, S. (2013). Chinese Guanxi: An integrative review and new directions for future research. *Management and Organization Review, 9*(1), 167–207.

CN.Chinadaily. (2021). *Ministry of Culture and Tourism in China: From 2004 to 2019, the number of people who participated in Red tourism increased from 140 million to 1.41 billion*. 23 March 2021, http://cn.chinadaily.com.cn/a/202103/23/WS605ada9ba3101e7ce9745945.html

Coles, T. (2003). Urban tourism, place promotion and economic restructuring: The case of post-socialist Leipzig. *Tourism Geographies, 5*(2), 190–219.

Dingli, A., & Seychell, D. (2015). *The new digital natives: Cutting the chord*. Springer.

Fei, X. T. (1992/1947). *From the soil: The foundations of Chinese society*. University of California Press.

Gao, Z., & Guo, X. (2017). Consuming revolution: The politics of red tourism in China. *Journal of Macromarketing, 37*(3), 240–254.

Gu, H., Ryan, C., & Zhang, W. (2007). Jinggangshan Mountain: A paradigm of China's red tourism. In C. Ryan (Ed.), *Battlefield tourism: History, place and interpretation* (pp. 59–65). Elsevier.

Guo, J. (2022). The postfeminist entrepreneurial self and the platformisation of labour: A case study of yesheng female lifestyle bloggers on Xiaohongshu. *Global Media and China, 7*(3), 303–318.

Han, Z. (2015). Party building in urban business districts: Organizational adaptation of the Chinese Communist Party. *Journal of Contemporary China, 24*(94), 644–664.

Henderson, J. C. (2007). Communism, heritage and tourism in East Asia. *International Journal of Heritage Studies, 13*(3), 240–254.

Hope, M. (2009). The importance of direct experience: A philosophical defence of fieldwork in human geography. *Journal of Geography in Higher Education, 33*(2), 169–182.

Hung, C. T. (2018). Communist tradition and market forces: Red tourism and politics in contemporary China. *Journal of Contemporary China, 27*(114), 902–923.

Koss, D. (2021). Party building as institutional bricolage: Asserting authority at the business frontier. *The China Quarterly, 248*(S1), 222–243.

Li, Y., & Hu, Z. (2008). Red tourism in China. *Journal of China Tourism Research, 4*(2), 156–171.

Li, Y., Hu, Z., & Zhang, C. (2010). Red tourism: Sustaining communist identity in a rapidly changing China. *Journal of Tourism and Cultural Change, 8*(1–2), 101–119.

Light, D. (2001). "Facing the future": Tourism and identity-building in post-socialist Romania. *Political Geography, 20*(8), 1053–1074.

Liu, C. (2021). Exploring selfie practices and their geographies in the digital society. *The Geographical Journal, 187*, 240–252.

Liu, C. (2022). Imag(in)ing place: Reframing photography practices and affective social media platforms. *Geoforum, 129*, 172–180.

Luo, X., & Bao, J. (2019). Exploring the impacts of tourism on the livelihoods of local poor: The role of local government and major investors. *Journal of Sustainable Tourism, 27*(3), 344–359.

Marvell, A., & Simm, D. (2018). Emotional geographies experienced during international fieldwork: An evaluation of teaching and learning strategies for reflective assessment. *Journal of Geography in Higher Education, 42*(4), 515–530.

Nairn, K. (2005). The problems of utilizing "direct experience" in geography education. *Journal of Geography in Higher Education, 29*(2), 293–309.

National Red Tourism Coordination Executive Team. (2005–2015). *Annual reports on the Chinese red tourism development*. China Tourism Publishing House.

Pan, S. Y., Gao, M., Kim, H., Shah, K. J., Pei, S. L., & Chiang, P. C. (2018). Advances and challenges in sustainable tourism toward a green economy. *Science of the Total Environment, 635*, 452–469.

Perrin, A. J., & Gillis, A. (2019). How college makes citizens: Higher education experiences and political engagement. Socius: Sociological research for a dynamic. *World, 5*, 1–16.

Pinggu Statistical Bureau. (2019–2022). Statistical yearbook of Pinggu District. .

Tang, W., Zhang, L., & Yang, Y. (2021). Can red tourism construct red memories? Evidence from tourists at mount Jinggang, China. *Journal of Destination Marketing & Management, 20*, 100618.

Timothy, D. J. (2007). Introduction: The political nature of cultural heritage and tourism. In D. J. Timothy (Ed.), *The political nature of cultural heritage and tourism: Critical essays* (Vol. 3, pp. ix–xviii). Ashgate.

Timothy, D. J. (2011). *Cultural heritage and tourism: An introduction*. Channel View.

UNWTO. (2018). *Statistical framework for measuring the sustainability of tourism*. World Tourism Organization. https://webunwto.s3.eu-west-1.amazonaws.com/s3fs-public/2020-09/wge_mst_2nd_item_2.1_doc_0.pdf

Wang, J. W. (2023). *Annual report on the development of red tourism in China (2022)*. Social Sciences Academic Press.

Wang, H., Zhang, B., & Qiu, H. (2022). How a hierarchical governance structure influences cultural heritage destination sustainability: A context of red tourism in China. *Journal of Hospitality and Tourism Management, 50*, 421–432.

Weber, M. (1978). *Economy and society* (2nd ed.). University of California Press.

Weibo. (2022). *Immersive Experience Activity of Red Party Building*. https://weibo.com/ttarticle/p/show?id=2309404796186090668565

Williams, R. (1977). *Marxism and literature*. Oxford University Press.

Yan, X., & Huang, J. (2017). Navigating unknown waters: The Chinese Communist Party's new presence in the private sector. *The China Review, 17*(2), 37–63.

Zhao, S. N., & Timothy, D. J. (2015). Governance of red tourism in China: Perspectives on power and guanxi. *Tourism Management, 46*, 489–500.

Zuo, B., Huang, S., & Liu, L. (2016). Tourism as an agent of political socialisation. *International Journal of Tourism Research, 18*(2), 176–185.

Zuo, B., Gursoy, D., & Wall, G. (2017). Residents' support for red tourism in China: The moderating effect of central government. *Annals of Tourism Research, 64*, 51–63.

Ran Liu is an Associate Professor at the College of Resource Environment and Tourism, Capital Normal University, Beijing, China. She holds a PhD in Human Geography from Nanyang Technological University (Singapore). Her research interests include the urban village (*Chengzhongcun*), informality, spatial politics, land and housing policies, and migration issues in transitional China. To date, she has published in international refereed journals (including Environment and Planning A, Cities, Habitat International, Applied Geography, Urban Policy and Research, Asia Pacific Viewpoint, etc.), the monograph *Spatial Mobility of Migrant Workers in Beijing, China* (Springer 2015) and *Urban Village Redevelopment in Beijing, China: New Housing Opportunities for Migrant Workers* (Springer 2024), and chapters in the book *The Routledge Handbook of Henri Lefebvre, the City and Urban Society* (Routledge 2020) and *Population Mobility, Urban Planning and Management in China* (Springer 2015).

Chapter 10
Thinking About Ethnic Minority's Rural Festival Cultural Tourism: A Case of Guangxi, China

Xiangzhu Lei

1 Introduction

The festivals of ethnic minorities embody the core culture of each ethnic group, as they generally represent important seasonal or significant events. Despite regional and lifestyle differences between ethnic groups who still maintain their unique festivals, some festivals are shared among them due to increased interactions.

Of those unique ethnic festivals in Guangxi province, the Zhuang's "Sanyuesan", the Jing's "Ha Jie", the Yao's "Panwang Jie", "Zhuzhu Jie," and "Shaiyi Jie," and the Mulao's "Yifan Jie" stand out most distinctively in attracting numerous visitors from all directions. These unique festivals are closely connected to their local origins and traditional elements, and they have enriched themselves respectively with their own ethnic minority cultural features which are highly valuable in terms of rural tourism quality. In recent years, ethnic minority festival tourism in Yunnan province has developed and gained an increasing attention from local governments and the tourism industry. However, during my recent travels around in the Yunnan regions, I have discovered a phenomenon of homogenization and superficiality in their festival tourism products. Moreover, as festivals are increasingly commercialized to pursue economic benefits, it poses significant challenges in preserving the quality the ethnic traditional culture. This is particularly noticeable in some places where government-built ethnic villages overlook the authenticity of ethnic culture which by right should be closely associated with its original cultural ecological spaces.

X. Lei (✉)
Guanxi Minzhu University, Naning City, China

T.-C. Wong et al. (eds.), *Cultural Tourism in the Asia Pacific*,
https://doi.org/10.1007/978-3-031-63459-8_10

As a starting point in our investigation on the loss of cultural authenticity, this paper takes the important festivals celebrated by some indigenous ethnic groups in Guangxi where we explore how best to balance between rural tourism promotion with cultural charm and the preservation of the authenticity of ethnic festivals. By promoting our outstanding ethnic minority culture, we aim equally to enrich the spiritual life of rural communities and develop more effectively the overall rural tourism.

2 Types of Ethnic Minority Festival Tourism

Cultural festivals are also known as culture-based holidays. They are usually referred to a nationwide cultural festival or those of ethnic minority groups which have been established over a long historical process encompassing features of specific customs and practices. Festivals represent the most distinctive and meaningful cultural symbols of a nation and they embody spiritual beliefs of the people nurtured with ethnic sentiments. In this aspect, ethnic groups in Guangxi have been recognized as having richly significant festivals classified at the national-level as intangible cultural heritage. Cultural festivals are also occasions where different ethnic minority groups come together to experience and share the sense of sacredness brought by unique rituals. Through the shared sensory and spiritual joy obtained from the festival celebrations, they enhance their interpersonal communication and social relationships, and strengthen their collective memories.

During festival activities, the specific aspects of ethnic minority culture, such as religious beliefs, ideologies, traditional attire, language, customs, cuisine, and handicrafts, are expressed and showcased in a concrete, dynamic, and centralized manner, making them highly valuable cultural experiences (Tan et al., 2020). The festival venues have played an important role in bringing together different ethnic groups to interact and as an authentic public space in sharing enjoyable moments.

Indeed, ethnic festivals, as tourism resources, constitute a source of cultural tourism with strong appeal and a high level of interactions between hosts and guests, offering deep and memorable experiences. They have practical significance for tourism economic development, promoting exchanges among various ethnic groups, and preserving the heritage of ethnic cultures (Wang, 2019). They enable individuals of different ethnic groups to understand and respect cultural practices of each other. It is a common practice that during ethnic festivals, leaders of one ethnic group would invite participants from neighbouring ethnic groups. As the hosts, they would showcase their hospitality, encourage their neighbours to take part in the entertainment activities. Activities where neighbours are commonly involved in sharing the fun are lantern exhibitions, dragon and lion dances, and folk parades. From my personal observations, the ethnic minority festivals in Guangxi can be roughly categorized into the following three types of tourism activities.

2.1 *Comprehensive Brands of Festival Celebrations*

The "Sanyuesan" festival is a traditional festival celebrated jointly by the Zhuang, Yao, Miao, and Dong minority groups in Guangxi, among whom the Zhuangs are most prominent participants. With a history of over a thousand years, "Sanyuesan" has been able to endure and flourish because it is deeply rooted in the vast rural areas of various ethnic minorities where they organize the event every year. In 2014, the festival was officially designated as a public holiday festival in Guangxi which provides residents with a strong sense of ethnic significance and identity.

In 2017, this festival successfully evolved into a broader tourism brand known as the "San Yue San, Ba Gui Jia Nian Hua". In the following year, Guangxi province received more than 16.35 million visitors, generating a revenue of over 12.6 billion Yuan. The "San Yue San" celebration was held in large numbers of ethnic minority villages where a total of 640 songs and dances were performed, among 300 other ethnic-based cultural activities across the province (Guangxi Daily, 2018).

This comprehensive brand festival has produced substantial impact in the Guangxi ethnic minority regions. The Dong and Miao ethnic groups in Chongzuo, for instance, showcased their performance such as the "Da Lang Dance," "Bronze Coin Dance," "Hua Shan Fist," and "Bamboo Pole Dance". Other ethnic regions like Liuzhou focused on folk song performances. In Sanjiang, they sang "Dong Ethnic Grand Song", and in Luzhai and Rongshui, it was the "Ping Shan Folk Song" and the "Three-Defense Song" they sang. Additionally, in Qinzhou, the fishermen performed the "Flute Music Festival", attracting tens of thousands of visitors (Luo, 2019).

As a matter of fact, the "San Yue San" festival has generated tremendous impact in the promotion of local tourism. For example, in early 2023, Baise City launched a "one-card" offer during the "San Yue San" festival, a special ticket of merely 198 yuan allowing visitors to enjoy sightseeing all A-class scenic spots in the city. Besides, visitors who chose to visit major attractions in the city were offered gifts of ethnic-designed souvenirs, free or discounted tickets for specific high-quality tourism routes such as the "Red Culture Tour," "Health and Wellness Residence Tour," "Intangible Cultural Heritage Experience Tour", "Folk Customs and Traditions Tour", and "Exploratory Tour to the Border Regions" (Liu, 2023).

2.2 *Scenic Spots and Cultural Festival Tourism*

This package refers to visits to renowned scenic spots in regions inhabited by ethnic minorities supplemented by cultural festivals with the objective to enhance their visibility and sphere of influence. Typical examples of this category are represented by the Longji Ricefield Terraces in Longsheng and the "Golden Beach" of the Jing ethnic group in Dongxing.

The Longji Ricefield Terraces are a 4A-National class scenic area and home to the Zhuang and Yao ethnic groups who had created the “Ping An Zhuang Terraces” and the “Jinkeng Hong Yao Terraces” respectively. Since 1992, these two sites have attracted numerous visitors with their natural scenery and cultural characteristics. Apart from the “San Yue San” festival, which celebrates the onset of the rainy season, the “Planting Festival” and “Transplanting Festival” reflecting well the Zhuang’s agricultural culture, are especially favoured by tourists. During the Planting Festival when local farmers work collectively in the rice fields, such a scenario would leave a strong visual impact on urban visitors. The Sun-drying Festival is another significant event whereby the Red Yao people of Longsheng would celebrate. It is on every 6th day of June, all households would display vividly their embroidered and colourful clothes and skirts under the sun throughout the villages. On this day, other activities like “Lifting the Golden Dog” and “Holding the Long-table Banquet” are also held, infusing the Longji Ricefield Terraces festival with richer cultural substance.

Situated in the southeastern corner of Dongxing City in Fangchenggang, Jingzu Sand Island’s tourism industry began to thrive in the 1990s. With a 10-km golden sandy beach, the island has become a well-known waterfront tourist destination. The peak tourist season to the Jingzu Sand Island falls on the celebration of the “Ha Jie” festival by the Jing ethnic group. This festival is a combination of religious rituals, ancestor worship, communal feast, and ethnic-cultural entertainments. As the grandest traditional cultural festival of the Jing ethnic group, there are a series of ceremonial rituals that last for 7 days, organized in four stages: (a) Welcoming the Sea God; (b) Offering sacrifices to the Sea God; (c) Singing the “Ha” song to entertain the Sea God; and (d) Bidding farewell to the Sea God (Zhong & Liu, 2019).

Apart from attracting other ethnic groups to visit the Jingzu Sand Island and to have a deeper understanding of the Jing culture, the “Ha Jie” festival offers a family reunion opportunity of Jing people from other regions to return to their home villages. As the sacred rituals and joyful singing activities take place, the whole ethnic community and the visitors turn themselves into participants and promoters of the festival (Fig. 10.1).

2.3 *Ethnic Festivals with Unique Influence*

As said, ethnic festivals originate from long-standing religious folk beliefs of individual groups and put to practice yearly in local areas with celebratory events. In the mountainous areas of Guangxi province, there are unique (jie) festivals of different ethnic groups such as the Yao’s “Panwang Jie” and “Zhuzhu Jie,” the Maonan’s “Fenlong Jie,” the Mulao’s “Yifan Jie,” and the Yi’s “Huoba Jie.” These festivals attract the interest of culture lovers to visit the region because of their unique cultural appeal.

Fig. 10.1 Ethnic minority festival celebrations in Guangxi Province (**a**) Mulao's Yifan ethnic festival worshiping 36 deities at Luocheng Mulao County town (**b**) Ritual ceremony in honour of Deity Panwang at Dongshan Yao Twonship, Quanzhou county (**c**) Ethnic Jing's Ha deity welcoming festival at Jin Beach, Jiangping Towbship (**d**) Ethnic Jing's Ha deity ritual ceremony at Jiangping Township. (Source: Photos by author taken during 2019–2023)

During the 1950s to the 1970s, ethnic cultural performances were virtually silent for nearly 30 years in Guangxi due to historical reasons. Following China's economic reforms after the 1980s, ethnic minority groups started to resume with vigour their traditional cultural practices. Important festivals were restored including folklore festivals and village-based celebrations with ritual ceremonies. From the perspective of tourism, the scope of celebrations has also expanded from localized scale before the advent of the twenty-first century with mainly local participants to large crowds involving local and nation-wide participants after the twenty-first century. Soon, the dynamics of the ethnic-led celebrations had captured the attention of the UNESCO (United Nations Educational, Scientific and Cultural Organization) whose role is to protect and safeguard the cultural diversities of human societies under its "Convention for the Safeguarding of the Intangible Cultural Heritage".

In 2003, UNESCO, in response to China's application, approved and accepted Guangxi's large-scale festivals of various ethnic minorities as intangible cultural heritage. UNESCO's recognition has bolstered substantially ethnic minority festival tourism in the province including the Nongdong Festival of the small Boudai ethnic group (a branch of the Zhuang ethnic group) in the town of Jinlong, Longzhou.

Other major festivals of the larger ethnic groups, such as the Panwang Festival of the Yao ethnic group and the Yi Rice Festival of the Mulao ethnic group, have been organized as cultural tourism festivals under the leadership of the government which provides various forms of publicity support.

Taking the Panwang Jie of the Yao people as an example, it originated from the Panwang Festival singing contest. Every October 16th in the lunar calendar, Yao people gather to sing and dance, commemorating their common ancestor Panwang. Over time, it evolved into the Panwang Jie, also known as "Huan Pan Wang Yuan," with celebrations taking place either by individuals in their homes or by the entire village. The ceremony is divided into two parts. The first part is the sacrificial ritual, which includes processes like "invoking the saints, arranging positions, lighting lamps, calling the harvest, fulfilling vows, and thanking the saints" (Liao, 2009).

The second part is the invitation of ancestral gods and the whole community would come together to enjoy the ceremonial rituals. The event lasts for either 3 or 7 days during which people sing the Panwang epic, listen to the teachings of Panwang, thereby strengthening their emotional attachment and sharing their common ethnic identity. Starting in 1990, with the government's support and initiation from the Guangxi Yao Society, a number of Yao autonomous counties jointly organized the Panwang Festival, with each county taking turn subsequently to organize once in every two years. The event was called the "7th China South Mountain Yao Panwang Festival" and was successfully officially recorded as "China Yao Panwang Festival"in 2006 (Tan, 2009).

Since 2001, the Panwang Jie has been integrated into cultural tourism festival events, where numerous song and dance performances, folk art exhibitions, sports events, trade and business negotiations, as well as product promotions are jointly organized. Similarly, in 2009, the Mulao's Yifan Jie was sponsored by the government of Luocheng Mulao Autonomous County, which was the first county-level Yifan Jie commemorative event. This festival which used to be a folklore celebration focused on expressing gratitude to heaven for a good harvest and praying for peace, has now become a grand festival integrating folklore, art, sports, trade, and tourism.

3 Bottlenecks in the Development of Ethnic Festival Tourism

From the brief discussion of festival tourism in Guangxi mentioned above, it can be observed that major ethnic festivals have gained widespread recognition due to government involvement and integration with tourism. These festivals are increasingly sought after by outsiders because of their distinct cultural characteristics and symbolic significance, fulfilling the need for experiencing exotic and diverse cultures. Take the Maonan's "Yifan Jie" as an example, since its inception in 2009 and its revised edition in 2013, the number of tourists has significantly increased, reaching a total of 460,000 visitors throughout 2013 (Pan & Zhou, 2014). Ethnic festivals indeed have demonstrated a notable increase in consumer spending, they have played an essential role in enhancing ethnic cultural identity, promoting social interactions, bringing together various ethnic groups, and making them a vital channel for urban visitors to understand better features of the ethnic minorities and have interactions with them as well.

Admittedly, while we can see the economic effectiveness of festival tourism, we also need to rethink that this kind of festival tourism promoted by the government and local communities has faced certain challenges in the area of long-term protection of ethnic heritage and culture and the sustainable development of ethnic rural tourism. In addition, due to the excessive commercialization and pursuit of economic benefits in ethnic tourism, the problems and bottleneck must not be underestimated. In this respect, the following issues have to be looked into carefully.

3.1 *Homogenization and Convergence: Uniformity in Festival Formats*

Government-led ethnic-cultural tourism-based festivals often aim to accommodate large audiences. They are commonly organized for large-scale performances in the form of folk parades, activities characterized by ethnic sports, feasts, and product promotions. At the surface level, such events may appear diverse and colourful. But their organizational format and marketing strategies often lack innovation, resulting in a shortage of unique festival memorabilia and handicraft-style skills.

Another issue which arises is that government-led celebrations usually need to attract audiences within a short span of time, with emphasis placed on the entertainment factor, and love narrative of the young, ethnic dancing and music performance often dominate a significant proportion of the whole event. For some festivals, even if the aim is to showcase the original essence of the ethnic culture, it has turned out difficult to achieve. For instance, in the performance of Maonan's "Yifan Jie", the ritual blessing ceremony was a simplified version without authenticity due to physical constraints on stage. As a result, the performance has incorporated a set of modern-day elements, giving the impression of a carnival-like atmosphere.

What was once a mysterious, solemn, and culturally distinctive ritual has been transformed into a shallow performance on public stage. The audience watches the spectacle but they could not share the authentic cultural feeling from the ritual. Many tourists expressed their disappointment, stating, "all the ethnic festivals look similar, lacking distinctiveness. Such a performance is merely the gathering of a big crowd and there is nothing special and unique. It's hard to find the genuine traditional cultural traits we have come to see". This sentiment represents the voices of travelers, primarily due to the fact that the performance lacks an in-depth study and artistic expression of the authentic traditional cultural festivals.

3.2 *Overemphasis on Commercialization: Diluting the Authenticity of Ethnic Festivals*

Reflecting originality is the core value of cultural heritage, and for minority festivals, originality means uniqueness, differentiation, and also an experience from their unique atmosphere, and specific stage designs, among others (Axelsen & Swan, 2010). Ethnic festivals are important intangible cultural heritage which can hardly be replaced by simple reproduction of traditional cultural festivals from which revenues are made (Getz, 1998). Intangible cultural heritage may have no written records. It may rely on people's words of mouth, and it has the characteristics of living fluidity, territoriality and vulnerability. Given its sensitivity, placing overwhelmingly emphasis on commercialized and large-scale development of festival tourism without proper planning, the authenticity of festival performance will have to suffer (Axelsen & Swan, 2010). The consequence may even have a damaging impact on the festival heritage, in which case its genuine value is difficult to be restored.

In Guangxi, some ethnic festivals had exhibited in the initial stage respect and faithful to their traditional ceremonial practices, while incorporating innovative elements. This was true of the first four versions of the Panwang Jie festival performed before 2001 where narratives were centred on cultural rituals, festivals and celebrations. However, as time went on, especially when business was taken over by tourism enterprises whose aim was more profit-oriented, the loss of festivals' authentic cultural content became increasingly evident. However, starting in 2001, the festival performing ceremonies have been gradually "hijacked" to serve business interests in attracting investment, promoting local cultural products and attracting tourists (Tan, 2009). For instance, during the 2012 Panwang Jie festival, an event called the "Yao Wedding Ceremony" was organized to showcase a scene of Yao brides weeping with tears during the marriage ceremony. In some areas, however, a "Yao Princess Contest" was held instead (Mo, 2012) which turned it into a beauty pageant. Such a modified episode was planned primarily to cater to the needs of investment companies aiming at maximizing economic returns through media publicity.

This kind of commercialization involves a vast number of intermediaries and organizations who are stakeholders in having an interest in satisfying their own business functions and goals. They have all in one way or another used the traditional heritage as "consumer goods". The loss of cultural essence in such events has turned performances into mere spectacles without substantial spiritual content. Consequently, they fail to arouse the interest of tourists in exploring exotic cultures, foster cultural identification, and casting doubts about the authenticity of their own beliefs. In the long run, this approach hinders the sustainable development of ethnic cultural festivals and their related economic activities (Peng, 2005).

Protecting the originality of national cultural festivals does not mean that they cannot further develop or evolve. In reality, cultural content and performance style are subject to change over time and space. In the process of developing national culture tourism, conceptual perceptions of tourists and local performers will affect each other in the interactive process. But the connotation of the culture of each ethnic group is the soul and core value in the competitiveness of rural tourism in ethnic minority areas. The fact is currently, exhibition activities held by tourism companies have gradually lost the essence of ethnic festivals by turning their valuable living cultures into an instrument of making money, resulting in the hollowing the genuine content of festivals with no substantive spiritual connotation. In the long run, such a practice is also not conducive to the sustainable development of the national cultural festivals.

3.3 Ethnic Festival Tourism, Failure to "Bleed" Ethnic Festival Heritage Sites

Ethnic cultural festivals originate from and thrive within the communities. They can maintain their cultural essence and be passed down to future generations only when deeply rooted in the fertile soil of the ethnic communities. It is evident that the integration point of ethnic cultural heritage and tourism lies not in county towns but in rural areas. Only when ethnic festival heritage sites in rural areas thrive can ethnic festival tourism be sustainable. However, at present, governments often focus on "providing a cultural stage" for economically viable performances, by concentrating on annual celebratory events or constructing some ethnic-themed parks in convenient locations near county towns.

These parks have been selected in places where local residents had left and they are transformed into purely commercial premises. This choice of sites near county towns instead of rural areas has indicated a missing link between ethnic festivals which are of rural origin with which there is a profound emotional connection and in ensuring the authenticity and attractiveness of ethnic festivals. At present, most local governments pay little attention to "blood transfusion" to ethnic heritage sites, mainly in the following ways:

They ignore the importance of discovering and protecting the "static" culture of local ethnic groups in two aspects: Preservation of ancient ethnic village buildings, agricultural landscapes formed by long years of production and farm labour, and the display of traditional agricultural production and life through the medium of physical objects including pictures, song books, scriptures and other media in museums or showrooms.

Traditional farm tools are often discarded by farmers because they are useless and take up space. Such houses with rich ethnic characteristics of the ethnic Zhuang are rarely seen. Only a few ethnic villages attach importance to the establishment of museums or exhibition halls to collect these traditional objects, such as Jinxiu Yao Autonomous County where there are a number of village museums, such as the Huarangyao Museum in Mentou Village.

(a) At present, in Guangxi, only a few villages can get the "Beautiful Countryside" development funds from the government. Most villagers nevertheless tend to demolish old houses and build new houses. In my field study, I found out the Napo black Zhuang ancient houses with a unique dry-rail type building style had disappeared.

(b) Little attention is paid to the training of young people to inherit traditional cultural festivals. It is important to take the young people to the heritage sites and teach them the past traditions some of which are only passed on from mouth to mouth from generation to generation. The living heritage mainly relies on the young people of the ethnic groups.

(c) With the development of society and the influence of the main stream dominant groups and urban culture, ethnic youths aspire more and more to live in the cities and know less and less their own ethnic cultures, making it even more difficult for them to become inheritors of ethnic traditions. It is common to discover that some local adults, even when are engaged in local ethnic tourism after their return from the cities, they lack a good understanding of their own culture. Consequently, it has become difficult for them to design or innovate brand-new cultural festival activities. Thus, it is crucial to train them from young and enable them to undertake and operate tourism business operations with decent heritage values in their mind.

4 How Ethnic Minority Festival Culture Can Better Lead Rural Cultural Tourism

Ethnic minority festivals are intangible cultural heritage, how to play its due role in rural tourism, and then better lead rural cultural tourism development is a topic worthy of in-depth discussion. For ethnic minority festivals to be sustainable in the long-term, their homes must be maintained with a cultural symbiosis, which has good potential to satisfy tourists' experiences as visitors. Robertson (2017:25) defines sustainability as something which endures "into the long long-term future; it refers to systems and processes that are able to operate and persist on their own over long periods of time".

In ethnic minority areas of Guangxi, the protection and utilization of heritage via tourism provides an important opportunity in sustaining local income and employment creation. Supporting tourism business as a form of knowledge economy and an educational resource, these activities stimulate creativity, develop cultural capital in the form of high-quality and original products and services. However, such heritage tourism development should consider not only its direct economic impact but also its values of ecological, social and environmental sustainability derived from the human capacity in institutionally creating conditions to help our societies to survive (Platje, 2011).

4.1 The Protection and Development of Ethnic Festivals Should Be Carried Out in the Way of "Living Inheritance"

"Living inheritance"refers to the protection of the intangible cultural heritage from which it has originated, grown and further develop to face real life challenges. Protecting the living inheritance is conducive to maintaining the vitality of intangible culture, a key element emphasized by the United Nations Convention for the Safeguarding of Intangible Cultural Heritage. People's extensive participation in heritage practices helps to revitalize cultural heritage is critical in making it a part of their daily life and this requires the government to shift its "visible hand" to the "invisible hand" of the grass-roots people (Zhang, 2019, Mu et al., 2010: 73). This means big ethnic festival activities are needed to be planned with government guidance provided to people to demonstrate their cultural values with their full participation as main players, known as the "return of custom to the people".

The Ha Festival of the Jing ethnic group is effectively very exemplary in this style of "return of custom to the people". In 2008 and 2009, the local government "upgraded" the Ha Festival to the "Dongxing Ha Festival" and "Fangchenggang Ha Festival" by organizing a grand ceremony with art show which were never as grand before (Lv, 2013). However, this official grand ceremony had overlooked the spiritual rituals inherent in these two festivals. In 2012, the local government reconsidered the mistake they made and they returned the responsibility to the Jing's Folk Affairs Committee to organize it to ensure the traditional ceremonial activities were orderly and solemnly followed. Though the scale of organization was smaller, their traditional rituals were well received by visitors, and this ceremonial style has continued ever since.

The Yao's Panwang Festival in Dongshan Village is another good example. Local government's grand sponsorship was dropped since 2013, and organization was returned to small villages in Dongshan where villagers were proud to hold their own festivals with fund collected from donations which would cover most of the cost while the local government would top up the difference. This is how it works now, showing the strong cohesive belief of the Panwang villagers.

From my field study in Dongshan, I witnessed with my own eyes the potential inheritors of the Yao's folk songs and dances in some villages. There were teachers conducting classes to train singers and dancers, among whom "singing king" and "dancing king" were elected. These teachers also teach in the primary schools, and they encourage the local children to love their own ethnic culture from an early age. With children as future inheritors with love and passion for their Yao culture, it is highly probable that with training, the traditional folk songs and dances will pass down successfully from generation to generation.

In this regard, Sun Zhengguo, a folklore researcher said thus: "These cultural inheritors are outstanding, they are representatives of an intangible culture and they will shape the path of the ethnic culture with their quality of virtue, art and skills.

Like cultural heroes, they play the leading role in three aspects, namely morally, skillfully and institutionally. They inherit their unique ethnic tradition, bring their people together, reconstruct their local knowledge system and provide a home with spiritual living, peace and prosperity" (Sun & Xiong, 2019).

4.2 *Promoting the Overall Development of the Ethnic Cultural Ecological Space*

Cultural ecological space refers to the state of cultural environment for human existence and development, and it forms a natural ecological space by itself. This cultural ecological space covers tangible and intangible domains, both of which are interdependent and intermingled. The basis of intangible cultural heritage protection is to maintain the continuity of inheritors and their cultural soil (Ma & Zhu, 2018). In order to inherit effectively and make full use of the leading festival cultural functions having its roots in rural tourism, more works should be done to ensure there is mutual support and harmonious development.

In response to this concern, we need first of all to protect the existing ancient village buildings in the heritage sites. These ancient villages embody architectural structures with strong local characteristics, providing traditional style in their outlook and reflecting specific features of certain historical periods. Such material components have a strong appeal to tourists and as such it is necessary that the government provide support in their repair and maintenance with professional guidance to retain their original style as much as possible.

Arguably, cultural performances require ample space to accommodate large crowds of visitors when big celebrations need to be organized. To resolve this issue, some countries have built "Cultural demonstration sites" for performance purposes (Fennell David, 2004). During my visit to Philadelphia of the United States some years ago, the presenters in the "Amish demonstration site" showed us various lively scenes of this distinctive European immigrant community. The demonstration site for visitors is not far from where the Amish live. Admission tickets were charged and the presentation materials showed many aspects of the Amish unique way of life and beliefs. Besides, visitors could purchase books and souvenirs about the Amish people, and they were also able to take a carriage ride around the village to get a more comprehensive feel about the Amish traditional life.

In Guangxi, there are also some ethnic villages where "Demonstration sites" have been built. For instance, the scenic spot "Yao Wang Tian Xia" in Lihu, Nandan is a demonstration site of the Baiku Yao. However, it should be noted that the construction of such demonstration site has not met the authenticity standard of that ethnic village. It is felt that the site should serve as a model for showing genuinely the traditional values of the local people. In this respect, the local government office and private companies involved should conduct serious research about the Yao culture before making any construction plans.

Secondly, other customary practices should not be ignored. These include folk songs, costumes, cutlery art, moral principles, marriage rituals which are important to be conserved and passed on. During our study tour in Dahua and Bama, we found that festivals encourage people to sing folk songs in both towns and villages where there are gathering places for people to come and sing monthly or weekly. Their love for folk songs has supported festival activities, and are a group performance event.

Thirdly, we need to build museums or folk objects exhibition centres. By incorporating ethnic cultural symbols, characteristics, celebrity stories, big ceremonial events into physical objects for show, visitors can have a deeper understanding of the historical background of the cultural festivals. Access facility and other relevant infrastructure to these centres should also be improved to meet the diversified needs of tourists.

4.3 Training of Ethnic Cultural Festival Inheritors

The cultivation of talents is an important factor to promote the tourism of ethnic festivals. The youths shoulder the mission of protecting and inheriting the culture of ethnic festivals. We can start with the two following measures:

The first is to upkeep the traditional mentorship. In China's long tradition of education, "father-son transmission" and "teacher-apprentice teaching" has always been regarded as a typical way of training and cultivating inheritors. Some scholars have pointed out that when passing down skills of intangible cultural heritage to the inheritors, you need to provide them with a living space in which their body is the main carrier of that inherited space physically and spiritually. That space will be filled up with appropriate knowledge, experience, skills, beliefs and customs. The larger the number of inheritors, the more will be the vitality of the heritage (Sun, 2020).

By tradition, the transfer of ethnic cultural festival practice is characterized by "memory, oral transmission and physical teaching", which need to be taught in person. Both teacher and student should build up an emotional bond through which skills training will be repeated again and again. Students will also learn how to lead a decent life philosophy, work attitude and professional ethics from teachers who provide mentorship. The government or cultural organizations should take measures to encourage the training of inheritors, by providing subsidies to young people who are willing to learn whilst rewards should also be given to successful teachers. Training workshops can be opened up to the public whereby interactions between students and tourists can take place and this will help in the learning process.

The second approach is to strengthen relevant education in local selected primary and secondary schools. These schools are a good source of inheritors where cultural festival skills training can be conducted and practiced. For example, students can be taught to appreciate cultural-historical sites, to learn stories and legends closely related to the festivals and to be guides for tourists in the ethnic culture museums. This will help students have a deeper understanding of their own ethnic culture. Those more dedicated to deeper learning can be nurtured to learn and practice the whole traditional rituals, and play folk musical instruments and performances.

4.4 Incorporate Modern Elements with an Open Mind and Let Tradition Shine Together with Modernity

In this modern era, protecting and inheriting ethnic cultural festivals should be incorporated into the way of life of our modern society. Doing so will make it easier for our younger generation to accept it as part of their own daily life. For instance, the traditional handicrafts of the Zhuang, Yao and Dong ethnic groups can be combined with modern design concepts to create products that maintain traditional characteristics yet conform to modern aesthetics. Such products can be better accepted by the market.

In Guangxi's Hezhou, Jinxiu and some other places, we have come across a group of knowledgeable ethnic costume inheritors. They have not only inherited the embroidery skills of their own ethnic costumes, but have also created innovatively costumes with styles favoured by younger people. Their work has been well received by the public, thus promoting not merely the local economy, but also rendering more people to love their ethnic cultural products. Amazingly, this group of young entrepreneurs are well versed with modern digital technology which they use in their innovative work.

Indeed, enterprises should be encouraged to employ creative talents to incorporate ethnic cultural elements into their design, advertising, tourism, games and other businesses. This will clearly help promote cultural interchange and enhance values of the marketable products. To achieve this objective, there is a need to encourage young people who have left their village home earlier to return and start their business there. Through preferential policies for talents, the local government has the capacity to attract local youths, especially those who have received higher-education in the cities to come home and undertake business that has the potential to promote ethnic cultural tourism and conserve their ethnic heritage.

In conclusion, the above proposed pathway has the prospect to help ethnic cultural festivals to play an important role in rural cultural tourism and yet retain the essence of the living inheritance of the ethnic minorities. Through training ethnic cultural talents, we have hope that the sustainable development of ethnic festival activities will be assured and kept vigorously alive. With more authentic and immersive ethnic cultural festivals to be performed and offered, the localized cultural tourism industry in China will not only be enriched but will also draw in more satisfying visitors.

References

Axelsen M, Swan T (2010) Designing festival experiences to influence visitor perceptions: The case of a wine and food festival.Journal of Travel Research 49(4): 436–450.

Fennell David, A. (2004). *ShengTai LvYou* (p. 70). Trans. Z. Lingyun. Tourism Education Press

Getz, D. (1998). Event tourism and the authenticity dilemma.In. In W. Theobald (Ed.), *Global tourism* (2nd ed., pp. 409–427). Heinemann.

Guangxi Daily. (2018). *"San Yue San festival Guangxi carnival" highlights abound – bringing joy to tourists and prosperity to the people*, 2018/04/27.

Liao, M. (2009). Panwang festival of Yao ethnic group. *Guangxi Ethnic Studies, 04*, 209–211.

Liu, Mingyue. (2023). People's media. Tourism income of 484 million yuan ignites the baise tourism market with "San Yue San" series of activities. *People's Daily* Online, 2023-4-28.

Luo, X. (2019). Study on the synergetic development of ethnic cultural industry and minority village tourism: Taking "san yue san festival Guangxi carnival" cultural tourism brand as an example. *Agricultural Economy, 6*, 79–81.

Lv, J. (2013). From "Celebrating with the people" to "returning to the people": A case study of Ha Jie of Jing ethnic group. *National Arts, 6*, 94–98.

Ma, G., & Zhu, W. (2018). *Cultural anthropology and intangible cultural heritage*. SDX Joint Publishing Company.

Mo, S. (2012). "Princess of Yao" adds color to the 12th Chinese Panwang Festival of Yao ethnic group in Hunan, Guangxi, and Guangdong provinces. *Minzu Tribune, 23*, 70.

Mu, Y., Tan, H., & Liu, Z. (2010). *Introduction to intangible cultural heritage*. Beijing Normal University Press.

Pan, G., & Zhou, H. (2014). Research on the development of ethnic cultural economy from the perspective of spatial vision: A case study of Yifan festival of Mulao ethnic group in Guangxi. *Reformation & Strategy, 30*(8), 88–91.

Peng, Z. (2005). Human heritage and homeland ecology. *The Ideological Front, 31*(6), 116–120.

Platje, J. (2011). *Institutional capital-creating capacity and capabilities for sustainable development*. Wydawnictwo Uniwersytetu Opolskiego.

Robertson, M. (2017). *Sustainability principles and practice*. Taylor & Francis.

Sun, F. (2020). The generation and development of the concept of "living heritage" of intangible cultural heritage. *Cultural Heritage, 3*, 35–41.

Sun, Z., & Xiong, J. (2019). A multidimensional genealogy of heirs of intangible cultural heritage from the perspective of local elite culture. *Journal of Hubei University for Nationalities* (Philosophy and Social Sciences), *2*,15–22.

Tan, H. (2009). Reflection on the practice of protecting intangible cultural heritage of ethnic minorities: A case study of Chinese Panwang festival of the Yao ethnic group. *Guangxi Ethnic Studies, 2*, 172–178.

Tan, Y., Kang, Z., & Zhong, H. (2020). Study on the positive interaction between festival tourism development and traditional culture protection in minority areas. *Modern Business, 20*, 43–44.

Wang, Z. (2019). Research on the influencing factors of ethnic festival tourism resource protection: A case study of "San Yue San Festival" in Guangxi. *Business & Economy, 6*, 67–68.

Zhang, C. (2019). Review of community participation in heritage practice. *Research on Heritages and Preservation, 2*, 34–40.

Zhong, X., & Liu, X. (2019). Ritual process, symbolic meaning, and practical enlightenment of Ha Jie of Jing ethnic group. *Guangxi's Local Records, 05*, 55–60.

Xiangzhu Lei, Professor at the School of Marxism, Guangxi University for Nationalities, previously Vice Dean of the School of Educational Science, is Chairperson of the Guangxi Lüyin Women and Children's Social Service Center and Vice President of the Guangxi Social Psychology Association. She holds a BEd, Guangxi Normal University, an MA in political thoughts education, Zhongshan University, and a PhD in law, Guanxi University of Nationalities. She conducts research on education and adolescent psychological development in minority areas focusing on the impact of culture. She led projects related to culture, including research on the impact of folk religious beliefs on ethnic education in Guangxi's border areas. She immersed herself in ethnic minority communities in Guangxi, participating in large-scale ethnic festivals, while investigating the origins and development of festival cultures. She published two monographs and over 60 papers. Her monograph, *Research on Yao Girls' Education from a Gender Perspective,* won Third Prize for Outstanding Research Achievements in National Educational Science.

Chapter 11
Promoting Lion Dance as an Intangible Cultural Heritage via Education in Lingnan, China

Xiaoyu Wang, Guo Zhen, Ziyi Qiu, and Mu Zhang

1 Introduction

Lion dance is a form of experiential teaching and has a curriculum which combines teaching and practical activity, and it thereby fully embodies the concept of the unity of knowledge and action. A Chinese saying goes that "Reading a million books is equally important as travelling a million miles". Since ancient times, travel has been a respected means of learning and intellectual growth. The Ming Dynasty geographer and traveller Xu Xiake travelled to the four corners of the country, covering more than 20 provinces and regions, studying history and geography and recording a variety of natural and humanistic phenomena along the way. Xu in China has been seen as a pioneer who had a strong classical scholarship backed up with practical knowledge enriched by study tours.

The growing importance of practical knowledge for tourism students was much highlighted by Tribe (2002) under the terminology 'philosophic practitioner' and by Dredge et al. (2012) in terms of 'phronesis'. In September 2017, the Chinese Ministry of Education released the "Guidelines for Comprehensive Practical Activity Programs for Primary and Secondary Schools", whereby practical curriculum has been defined as an interdisciplinary and practical teaching material with

X. Wang
School of Management, Jinan University, Guangzhou, China

G. Zhen · M. Zhang (✉)
Shenzhen Tourism College, Jinan University, Shenzhen, Guangdong Province, China
e-mail: zhangmu@jnu.edu.cn

Z. Qiu
Shenzhen Tourism College, Jinan University, Shenzhen, Guangdong Province, China
e-mail: qzyqzy23@stu2023.jnu.cn

T.-C. Wong et al. (eds.), *Cultural Tourism in the Asia Pacific*,
https://doi.org/10.1007/978-3-031-63459-8_11

which to train students with more rounded qualities. The program is organized into four major categories, namely the investigation and inquiry, social service, design and production, and vocational experience. Among them, practical curricula are included in the field of "investigation and inquiry" of the Integrated Practical Activities Program, and are thus an inseparable and important part of practical training.

Lion dance as a cultural heritage, its content of practical curriculum is indispensable. As both a tourism an educational resource, the Lingnan's Awakening Lion Dance which originated in Nanhai, Foshan of Guangdong province has been included in China's national intangible cultural heritage list. As lion represents the beast of prey, the lion dance symbolizes good luck, and in every festive event, especially the Chinese New Year, there are lion dance performances. This tradition has been preserved and spread widely throughout the country and overseas where there are ethnic Chinese people. The representative figure of the lion dance, Wong Fei Hung, who performed the art of the lion dance, won the title "King of Chinese Lions" in a lion dance competition in the late Qing Dynasty.

The Lingnan's Awakening Lion Dance symbolizes the national soul of the Chinese nation. Its curriculum requires students to be trained in three aspects: learning the fundamental movements of lion dance, about the culture and history of lion dance, inculcating in them their pride of this national heritage, and fostering their team spirit, perseverance, and hard work. Lion dance activities are widely spread in overseas Chinese communities. Where there are ethnic Chinese people overseas, lion dance is there. Lion dance has indeed become an important ceremonial event to promote China's national prestige symbolic of the soul of China, and it serves as a cultural bridge for overseas compatriots to recognize their ancestry asset and bond with their ancestors. Its cultural value and influence are far-reaching. During festive seasons, the Lingnan's Awakening Lion Dance tours are performed which constitute an effective way to pass on intangible cultural heritage and to promote students' physical and mental health via a comprehensive education program. Without doubt, the dance constitutes an important source of revenue for organizing institutions as it attracts large crowds of domestic and international tourists.

Our literature search has revealed that many studies offering analyses of factors affecting practical curriculum of lion dance, there is however a lack of empirical analysis on the thematic content of practical curriculum. This study's objective is to investigate the Lingnan's Awakening Lion Dance's practical curriculum as to how it could help to preserve this cultural heritage and how lion dance performance can be used as a means to promote tourism development. It collects the views of students, teachers, tourism enterprise managers, and the leaders of the lion club on practical curriculum development from a wide variety of organizations. Through participant observations, semi-structured interviews, and questionnaire survey, our study explores the factors influencing the future development of lion dance based on which we wish to recommend a practical curriculum to make the lion dance more sustainable and lively in both tourism development and as a cultural heritage.

2 Literature Review

2.1 *Practical Curriculum*

In teaching practice, the practical curriculum of a program is an interdisciplinary approach that uses real-life situations to set up class activities, and adopts empirical methods, experience and inquiry to train high quality students (Tian & Kong, 2022). Equipped with basic features of autonomy, openness, inquiry and integrated activity skills, a practical curriculum not only embodies the basic concept of real-life knowledge, but it also can reflect the characteristics and fulfil the roles of the curriculum appropriately (Zeng, 2022). In so doing, practical curriculum can enrich and develop students' contents knowledge and thus realize the practical goals and values.

At present, there is a lack of good understanding of the concept of "practical curriculum "in the academic world, and some researchers prefer to call it "educational tourism"or "study tours." Olga (2004) have provided certain basic meanings to a variety of study tours. Some others have applied experiential practices to imitate real-life practices, helped transfer theoretical notions into practical concepts, and assimilated content knowledge into real-life business contexts (Kim & Jeong, 2018). Practical knowledge and skills acquired in such circumstances can be further explored at a later stage (Zhang et al., 2021). For tourism students, focusing on "doing" through experiential learning has been proven useful in improving students' skills and practical initiatives (Doğantan, 2020; Guachalla & Gledhill, 2019; Özoğul et al., 2020).

2.2 *The Awakening Lion Dance*

The Awakening Lion Dance has its earliest written record in the Han Dynasty (202 BC–220 AD). Now, the Awakening Lion Dance is one of China's outstanding folk arts. As a national intangible cultural heritage, lion dance culture has been circulating in Guangdong for more than one thousand years, and it is performed in festivals and is meant good will for people to avoid misfortune and disasters. Each lion is generally performed by two people, one dances the head and one dances the tail. Under the gong and drum music, the performers dress up as a lion and imitate various forms of lion movements. In the course of the performance, the lion dancers would demonstrate very masculine acts of Guangdong's southern martial arts. Perceived as an auspicious relic for exorcising evil spirits and avoiding misfortune, lion dance is organized in festivals or a major event, and it has continued to flourish over the generations. The popularity of the lion dance had attracted the attention of other countries as well. McGuire (2019) aptly interpreted lion dance practice as a type of *kungfu*[1] training, suggesting that both the lion dance and *kungfu* are not

[1] Chinese martial art.

distinct from each other. Rather, lion dance could have been integrated as part of the larger umbrella of *kungfu* practice.

Away from China, Ying and Chiat (2016) discussed the evolution of the Chinese lion dance in Malaysia. Having inherited lion dance from China, local Chinese Malaysians have retained the tradition and continued to practice it and they have developed it innovatively and won once the championship in an international lion dance competition. The members of Gund Kwok, an all-women's dance group in Boston's Chinatown, also performed lion dance as a form of physical exercise and a means of self-defence (Avaunt, 2021; Xu et al., 2017). Though lion dance is flexible in form and can be practiced everywhere, the new style has become more demanding where dancers have to jump from pole to pole in the air, and perform other risky acts. With the increasing levels in difficulties, lion dance may incur all kinds of injuries in training and competition (Ling, 2019; Thompson et al., 2015).

Lion dance as a performing art has a number of values in humanism, entertainment, performance, sports and competitiveness. Zhou (2018) found that organizing lion dance tours in colleges and universities could improve students' physical and mental health, cultivate their sense of national pride, and awareness and desire to inherit this national culture. Cao (2020) argued that after the Guangdong Lion Dance was elected the first national intangible cultural heritage in 2006, the Lingnan region has begun to witness more martial arts practices, and lion dance has since been integrated a step further with martial arts which has good potential to improve the physical fitness of participating members and lead to the creation of a more unique form of performance.

Encouraged by the lion dance enthusiasm and its being an important intangible cultural heritage, the Guangdong provincial government released a series of film and television dramas "Huang Feihong" to promote the Lingnan Awakening Lion Dance. Several lion dance teams in the Guangdong region were invited to go overseas for exchange performances with overseas Chinese groups.

In summary, our emphasis is that lion dance must be integrated closely into the practical curriculum in the training of students. Thus, this paper explores the ways practical curriculum should be developed, and we will provide here a case study with an in-depth analysis.

3 Methodology

This study used a combination of qualitative and quantitative research methods. It included participant observation, semi-structured interviews, and questionnaires. First, through participant observation at a secondary tourism vocational school (hereinafter referred to as S Secondary Tourism School) in Guangdong Province, China where students undertook a 3-month internship and the authors of this study had benefited from their investigations in the Awakening Lion Dance practical curriculum offered by the school. Second, semi-structured interviews were conducted with eight teachers and students from the school, the managers of three local travel agencies, and a president of a local Lion Dance Club. Third, survey questionnaire forms were distributed to students in 22 secondary schools in Guangdong Province and Hong Kong and the data were subsequently processed and analyzed.

3.1 *Participant Observations and Interview Design*

During the three-month internship at S Secondary Tourism School from October 2020 to January 2021, our study team first examined the contents of Lingnan's Awakening Lion Dance practical curriculum. This program is actually a kind of extracurricular activity designed specifically for internship students to fulfil both practical-based and performance requirements.

The S Secondary Tourism School is located in Baiyun District, Guangzhou City, Guangdong Province, north of the picturesque South Lake national tourist resort and west of the national 5A (the highest level) tourist attraction Baiyun Mountain. In early 2016, the S School created the "Guangdong study tourism experience park" (hereinafter referred to as "experience park"), and in March 2017, the S School passed the assessment and became the first campus in Guangdong Province to be officially awarded the status as a national 3A-level tourism scenic spot. The school receives more than 10,000 overseas visitors.

The Lingnan's Awakening Lion Dance program offered by the School for over 5 years provides a typical feature of Lingnan's intangible cultural folk heritage. Till the present time, the program has trained hundreds of students and has received nearly 10,000 student visitors from other places at home and abroad.

For the survey, the interviewees were divided into four groups, namely: (a) two school teachers named hereinafter as T1 and T2; (b) school students, named as S1, S2, S3, S4, S5, and S6 thereafter; (c) three managers of travel agencies operating study tours in Guangzhou, named A1, A2 and A3; and (d) a president of a lion dance club in Guangzhou, named B1. The interviews with these two groups were about the travel agency experience and the views of the president of the lion dance club on the cultural heritage practical curriculum. The interviews mainly covered the interviewees' opinions on the market prospects for practical curriculum, the development of practical curriculum, and the influencing factors. More details are provided in Tables 11.1 and 11.2.

3.2 *Questionnaire Design*

Based on the literature review, the authors designed a questionnaire on perceptions of the market for the Lingnan's Awakening Lion Dance practical curriculum, with 38 questions borrowed from relevant established questionnaires from works of Bachner and Zeutschel (2009), Lee (2004), Liu and Chen (2020), McGladdery and

Table 11.1 Basic information of the interviewees in secondary tourism school S

Participants	Gender	Interviewee's age (years)	Interviewee's unit	Interviewee's position
Teacher T1	Male	43	S Tourism vocational and technical school	Teacher
Teacher T2	Male	37	S Tourism vocational and technical school	Director of teaching and research office
Student S1	Male	17	S Tourism vocational and technical school	Student (vice president of lion dance club)
Student S2	Male	17	S Tourism vocational and technical school	Student (lion dance club member)
Student S3	Male	15	S Tourism vocational and technical school	Student (not a member of the lion dance club)
Student S4	Male	16	S Tourism vocational and technical school	Student (not a member of the lion dance club)

Source: Authors' collation

Table 11.2 Basic information on practical curriculum of the interviewees

Participating interviewees	Gender	Working seniority (Years)	Interviewee's unit	Interviewee's position
A1	Male	20	(a) An international travel agency (b) A travel research and technology company	(a) Executive vice president (b) General manager
A2	Male	12	A private travel agency	Sales director
A3	Male	15	A state-owned travel agency	Sales director
B1	Male	8	A lion dance club in Guangzhou	President

Source: Authors' collation

Lubbe (2017), Olga (2004), Riegel and Kindermann (2016) and van Klooster et al. (2008). The questionnaires consisted of four parts as follows:

(a) Part 1: Demographic characteristics of the respondents;
(b) Part 2: General knowledge of the respondents about the Lingnan's Awakening Lion Dance program, their motives to learn and learning style. The first two sections consisted of 13 questions with multiple-choice (See Arcodia et al., 2021; Guachalla & Gledhill, 2019);
(c) Part 3: Respondents' perceptions of the value and development of practical curriculum (Cuffy et al., 2012; Petroman et al., 2016; Yang & Lau, 2019); and
(d) Part 4: Teaching and learning approach of the Lingnan's Awakening Lion Dance practical curriculum (Li & Liang, 2020; Stone & Petrick, 2013). The study team distributed questionnaires during the period January 4 to February 8, 2021, to the S Secondary Tourism School and other vocational school students.

There were a total of 22 vocational schools, among which eight from Guangzhou City, seven from Foshan and one from Jieyang. From Hong Kong, we received 26 valid responses from six schools.

3.3 *Trustworthiness*

This study adopted three different methods of data collection (in-depth interviews, participant observation, and focus group interviews). Multiple methods allowed for the triangulation of the data, with cross-checking between the different sources of information to improve accuracy and validity (Jick, 1979). All data collected from these methods were categorized and examined. The themes from the coding process were developed by assessing the information collected from different data sources. With multiple sources of information, bias could be minimized, and the trustworthiness of the results could be enhanced.

4 Results

4.1 *Observations and Interviews About the Lingnan's Awakening Lion Dance Practical Curriculum*

4.1.1 The Change of Inheritance Mode: From the Traditional Master-Apprentice Model to Vocational School Training

One of the authors took part in the teaching internship at S secondary tourism school where students have four sessions from 8:30 am to 12 noon and another two sessions from 2:00 to 3:25 pm, most of which are classroom-based classes such as Introduction to Tourism, Tour Guide Business, Hotel Management, English language and others. From 3:30 to 5:00 pm, students take extracurricular activities classes such as the Lingnan's Awakening Lion Dance and Tea Art. In the evening from 7:00 to 9:00 pm, it is self-study time after students have completed their homework and reading in the classroom.

The participating author visited the Lingnan's Awakening Lion Dance classes on two occasions where students were only those from the S secondary tourism school itself. Before the Lion Dance class begins, there is a general background briefing in the classroom for approximately 15 minutes on the history and characteristics of the intangible cultural heritage of the Lingnan's Awakening Lion Dance. The School expressed its great concern over the problem of getting young students to succeed and carry on the lion dance as an intangible cultural heritage. For this purpose, teaching focus is not merely on professional training of lion dance but also on the all-rounded quality of students.

From our investigation, we found that teachers and students in the School had executed the Lion Dance practical curriculum for more than 5 years. They had indeed a firm belief that the lion dance practical curriculum is critical in shaping their skills through campus training, and they would get stronger interest in lion dance and intellectual empowerment with a proud feeling in this rich intangible culture which they were prepared to carry the torch further on. A few reflections from the teachers and students are cited below:

> First, our school's practical curriculum has an extracurricular activity that we have created as a culturally unique feature. Second, our school campus is located at a Grade 3A scenic spot which has attracted the Hong Kong and Macau Reception Offices to use our school as a base to train their young students from Hong Kong and Macau. Also, we receive primary and secondary student visitors from the Hong Kong area and other areas of Guangdong Province to come and study here. (Teacher T1)

> Through this curriculum, the teacher explains to students the characteristics of Lingnan lion dance intangible culture, including the demeanour of the lion, its actions and movements, and the drumming with music. Dancing costumes and props are also used to show the decorated beauty of the stage and *kungfu*, and the movie "Wong Fei Hung" was played to inspire students to practice martial arts to strengthen their physical unity and endurance.

> The most important feature of the practical curriculum is not based on teacher lectures alone. Rather, it is the coaching that guides their training and help students build up their self-confidence, self-discipline and team spirit. Our school's current Lingnan lion team has succeeded from the previous students who now often go outside [the school] to perform. (Teacher T2)

> The recruitment of the Lingnan's Awakening Lion Dance practical curriculum is very effective, and every time we call for student registration, the class is fill up. Students who take up the program have a strong interest in it and they will learn to love this 'intangible' cultural heritage. (Teacher T1)

> I watched the movie 'Wong Fei Hung' and I admire him tremendously as a *kungfu* master. We have to perform lion dance in the school field, and we all train hard (Student S1). We are very demanding and strict in training, and we work together very well (Student S2). The Lingnan's Awakening Lion Dance practical curriculum can make you exercise, and strengthen your physical fitness. I have seen their performance and know that the school has a lion dance team. I am very interested and would like to sign up for it when I have the chance (Students S3 and S4).

4.1.2 The Dissemination and Promotion of Intangible Culture: Community Effects of the Lingnan's Awakening Lion Dance Activities

The research team also visited a lion dance club in a Guangzhou village and interviewed the club's president. The president said that the village had begun lion dance more than 100 years ago, with periods of ups and downs in between. This club has been established for 10 years and would organize at least one dance a month and it participates in commercial performances and in community events. The club which has more than 70 members who are primarily from nearby villages has won many awards and has conducted performances in Hong Kong, Macau, and Malaysia. It is believed that the club has shaped positively the village participants' character, and it is actively training primary and secondary school students to participate in the lion dance. The Club's President B1 had this to say:

> For primary and secondary school students who come in to learn the lion dance, we will ask them to respect their parents and study hard. In the lion dance competition, there is an emphasis on respectful and righteous spirit. Just like the practice of martial arts, we place emphasis on learning the righteous virtue first, followed by focus on study and promotion of physical health. I am very willing to go to schools to teach and pass on these priorities. Our routine training teaches students skills to do the proper movements of lion dance, and students have to learn how a lion would leap over high post spots, and express a lion's joy, anger, and sorrow. Our club has received invitations from some foreign organizations who request us to perform in a cultural exchange show, one of which is from Russia.

Scholars have continued to emphasize the value of a place-based educational approach rooted in local environments, consisting of a particular place's unique history, background, culture, economy, literature, and art (Howley et al., 2011). Whether it is from the interviews of teachers and students of S secondary vocational tourism school, or the questionnaire survey of 20 middle school students in Guangdong and Hong Kong, and the interviews of industry managers, the value of the course is unanimously recognized, which is a way of spreading excellent cultural heritage. In the questionnaire survey, the interviewees believed that the course could increase team cohesion, enhance sense of responsibility and self-confidence, and experience and share happiness. Travel agency managers mentioned that they will fully disseminate the theme of intangible cultural heritage when designing practical course products.

4.1.3 Marketing Prospects of Lion Dance in Tourism

The research team interviewed the managers of several travel agencies that conduct study tours.

Several managers expressed optimism about the market and tourism prospects for the Lingnan's Awakening Lion Dance practical curriculum conducted by S secondary tourism school. Travel agents have expressed their views as follows:

> If we choose the school Lingnan's Awakening Lion Dance practical curriculum products, we need to make a series of supporting actions, such as the characteristics of the lion from production to performance. While examining other study activities of the program, we are happy to cooperate with the school to develop and promote it. Practical curriculum for study tours involve schools, parents, and students, and attention needs to be paid to the combination and balance of the needs and feasibility of the three parties (A1).

> The agent A2 said that their company had been cooperating with this tourism school in the study practical activities, mainly by arranging students from other places to experience the curriculum of this tourism school. Several other travel agency managers agreed that schools are the primary decision-makers and critical actors in developing study tours as part of tourism activities. Nevertheless, the choice of practical curriculum for study tour destinations, the qualifications of the organizers, supporting infrastructure conditions, and prices are all

influencing factors. The travel agent A3 felt that the destination, the activity program, teachers' capability and skills are all more important factors.

From our field research, the interviewed travel agencies are generally optimistic about the support of national policies towards secondary school graduates who are interested to pursue this lion dance program. Some companies have also set up short-term practical courses in lion dance as part of their business expansion. As a tourism product, the concern is on cost-effectiveness, training quality among others. However, from our survey of students, we understand that middle school students have very limited knowledge of the practical courses of intangible cultural heritage such as lion dance. Despite the support of the national policy towards intangible cultural heritage elements, such practical-based programs are still in their infancy.

4.2 Introduction of the Essential Characteristics of the Questionnaire Sample

There were 104 male students and 205 female students among the questionnaire participants. There were more respondents from secondary tourism schools, which corresponded proportionally to the ratio of male to female students. The age structure and the structure of their grades matched, with mainly first- and second-grade students from vocational schools and first-grade students from public high schools, with most participants being aged 16–17 years old.

In their response, the participants' three main motives in taking up the lion dance programme which were knowledge acquisition, skills enhancement and interest fulfilment were recorded respectively at 61%, 58%, and 57%. Strangely, among the courses offered by the practical curriculum, close to half of the students expressed their strong interest in "tea brewing", and only 30% in the Lingnan lion dance, and the remaining 24% indicated their interest in bartending, ball sports, as well as etiquette, music, calligraphy.

Students were particularly interested in taking part in site visits away from their campus. Our survey showed when the Lingnan Lion Dance School organized a tour to visit Guangdong Provincial Tourism Experience Park in Guangzhou, a triple-A scenic spot, about two-thirds or 194 students of the whole group of 309 joined in while one-third were undecisive and only 4% or 13 did not participate. It has been found that tours that are closely linked to real life activities are particularly attractive to students. This has corresponded to the study carried out by Velempini and Martin (2019) whose findings showed that educational tours organized for high schools in Botswana linked to students' real life were much welcome by students. Real life pedagogic teaching is effective if it is linked to local histories and community activities.

4.3 *Factor Analysis of the Lingnan's Awakening Lion Dance Practical Curriculum Questionnaire*

In this study, SPSS 26.0 was used for exploratory factor analysis. Principal component analysis and maximum rotation were performed to extract factors with eigenvalues greater than 1. The correlation coefficients of each item and dimension were found to be more significant than 0.5 (0.604–0.804) for the three factor analysis. The cumulative variance explained was 68.29%, and the reliability was 0.91–0.93, meeting the reliability and validity criteria (Table 11.3). The three factors were as follows:

Factor 1: Lingnan's Awakening Lion Dance practical curriculum development;
Factor 2: Practical curriculum value; and.
Factor 3: Practical curriculum teaching style.

As shown in Table 11.4, the total score of the measurement index factors was 5. The mean values ranged from 3.58 to 4.19, with all indicators exceeding 3.5, indicating that all the measurement indexes had a high-value impact on students' learning. The standard deviation ranged from 0.852 to 1.114 which indicate minimal differences in respondents' opinions.

The highest mean value indicated that the largest problem of practical curriculum outside of the school was safety (4.19), which indicates that students' organization of outings has been dramatically restricted under the influence of the COVID-19 epidemic. This was followed by the fact that students could experience and share happiness through the program (4.11), that they could increase their interest in learning cultural knowledge through the program (4.07), that they could get out of the classroom and broaden their horizons through the program (4.06), that they could increase their practical exploration skills through the program (4.06), that they could increase team cohesion and collaboration (4.06), and that they could get along better with classmates through the program (4.06). In addition, the mean values for items indicating that having a teacher explain the cultural knowledge related to the Lingnan's Awakening Lion Dance is important (4.05), that there should be a professional lion dance team master to guide the training (4.02), that the course can increase self-confidence (4.02) and that the system can increase the sense of responsibility (4) were all above 4.

The lower mean values for the influencing factors included that the practical curriculum involved learning based on tourism (3.84) and that the practical curriculum should have assessment and evaluation (3.81). This indicates that students did not want practical curriculum to be just an assessment of going from one campus to another "campus" but rather wanted their experiences to be valued and did not want the practical curriculum for study tours to be mainly tourism.

Table 11.3 Awakening Lion Dance Exploratory factor analysis

Item	Factor 1	Factor 2	Factor 3
A1. The course will increase the ability for practical exploration	0.604		
A2. Through this course, students can get out of the classroom and broaden their horizons	0.65		
A3. Students can experience and find the course enjoyable	0.719		
A4. Through the course, students can enhance their sense of responsibility	0.685		
A5. The course can enhance one's self-confidence	0.735		
A6. Students can get along well with each other as classmates through this program	0.797		
A7. Students can learn with strong interest in cultural knowledge	0.749		
B1. Students will understand well the Lingnan Awakening Lion Dance practical curriculum		0.689	
B2. The program for the Lingnan awakening lion Dance will be taught in great details		0.767	
B3. A teacher leader introduces the lion dance program and its objective		0.744	
B5. A teacher leader is responsible to explain and demonstrate the standard movements of the lion dance		0.633	
B6. Students will be trained to master the essential elements of the lion dance		0.652	
B7. The student team members are trained to perform lion dance skills		0.634	
C1. A professional lion dance expert is assigned to supervise lion dance training			0.698
C2. A teacher should explain the cultural knowledge related to the Lingnan's awakening lion dance			0.747
C3. A visiting professional lion dance team should be invited for exchange purposes			0.771
C4. The greatest problem of practical curriculum of school is safety			0.678
C5. The practical curriculum should be taught in the same professional school curriculum			0.752
C6. The practical curriculum should include courses linked to tourism			0.756
C7. The practical curriculum provides opportunities to undertake exploratory learning trips			0.804
C8. The practical curriculum should have assessment and evaluation of students			0.622
Cumulative % (α)	54.61% (.93)	63.57%(.9)	68.29% (.91)

Source: Authors' computation

Table 11.4 Mean value analysis of practical curriculum development (N = 309)

Item	Mean	SD
A1. The course will increase the ability for practical exploration	4.06	0.926
A2. Through this course, students can get out of the classroom and broaden their horizons	4.06	0.96
A3. Students can experience and share happiness through this course	4.11	0.88
A4. Through the course, students can enhance their sense of responsibility	4	0.928
A5. The course can enhance self-confidence	4.02	0.92
A6. I can get along better with my classmates through this course	4.06	0.866
A7. I can increase my interest in learning cultural knowledge through this course	4.07	0.852
B1. The Lingnan's awakening lion Dance practical curriculum will be well-known among students	3.58	0.885
B2. There should be a detailed introduction or promotion of the program for the lion dance practical curriculum	3.72	0.958
B3. There should be a teacher (or leader) at the beginning of the course to introduce the purpose	3.88	0.864
B5. The teacher (or leader) should explain and demonstrate the standard movements during the course	3.72	1.114
B6. Through the course, training students can master the essential elements	3.93	1.046
B7. Through the course, team members can perform in some activities to show	3.88	0.959
C1. There should be a professional lion dance team master to guide the training	4.02	.873
C2. A teacher should explain the cultural knowledge related to the Lingnan's Awakening Lion Dance	4.05	.818
C3. There should be a professional lion dance team visit to learn and exchange	3.97	.827
C4. The greatest problem of the practical curriculum for study tours out of school is safety	4.19	.875
C5. The practical curriculum for study tours should be taught in the same professional school curriculum	3.97	.815
C6. The practical curriculum for study tours is a kind of learning based on tourism	3.84	.844
C7. The practical curriculum for study tours are mainly exploratory learning trips	3.98	.787
C8. The practical curriculum for study tours should have assessment and evaluation	3.81	.862

Source: Taken selectively from Bachner and Zeutschel (2009), Lee (2004), Liu and Chen (2020), McGladdery and Lubbe (2017), Olga (2004), Riegel and Kindermann (2016), van Klooster et al. (2008)

4.4 *Factor Regression Analysis of the Lingnan's Awakening Lion Dance Practical Curriculum*

Based on the above factor analysis, regression analysis was conducted on the variables represented by the three common factors, with the Lingnan's Awakening Lion Dance practical curriculum development as the dependent variable and the factor dimensions of curriculum value and curriculum teaching approach as the independent variables. The regression model was obtained, as shown in Table 11.5. When the F test was conducted on the model, it was found that the model passed the F test

Table 11.5 Regression model and correlation coefficients

	Non-standardized coefficient		Standardization coefficient						
	B	Standard error	Beta	t	p	VIEW	R^2	Adjust R^2	F
Constant	2.263	1.108	–	2.043	0.042*	–	0.726	0.724	F(404.69) p = 0.000
Practical curriculum value	0.635	0.034	0.747	18.755	0.000**	1.768			
Practical curriculum teaching style	0.163	0.044	0.148	3.728	0.000**	1.768			

Dependent variable: practical curriculum development factor, * p<0.05, ** p<0.01, *** indicates p< 0.001

Source: Computed by authors

($F = 404.69$, $p = 0.000 < 0.001$), indicating that the study's two independent variables, the value of the practical curriculum and the influencing factor of the teaching style of the practical curriculum, both positively influenced the development factor of the practical curriculum. Among them, the regression coefficient of the value of the practical curriculum was more significant and had a more substantial impact on the practical curriculum development, and the regression model equation was:

$$\textit{Practical curriculum development factor} = 2.263 + 0.635 \times \left(\text{practical curriculum } \textit{value}\right) + 0.163 \times \left(\text{practical curriculum } \textit{teaching style influence factor}\right).$$

5 Conclusions

Based on the case study of the practical curriculum about Lingnan Lion Awakening dance, this paper verifies the role of the practical curriculum on the dissemination of intangible cultural heritage, and it further explores the influencing factors of the development of its practical curriculum using a combination of qualitative and quantitative methods, such as participant observation, semi-structured interviews, and questionnaires, to corroborate each other and obtain the following points of conclusion.

(a) *Practical curriculum is an effective method of embodied learning outdoor*
Based on the authors' participatory observations, the Lingnan's Awakening Lion Dance practical curriculum is different from other general cultural courses in that students attend classes mainly outdoor where they are intellectually and physically trained with relevant skills. Outdoor open spaces with sunlight and fresh air

motivate students to pick up cognitive theoretical and practical skills. They are exposed to the outside world furnished with bodily composure filled with a rich spiritual soul, as argued in Kolb's experiential learning theory (Kolb & Kolb, 2009). It is based on this holistic model of learning derived from learners' own cognitive experience that a stronger learning outcome can be achieved (Davis et al., 2016; Girvan et al., 2016).

Our survey results have also shown that learning motivations in the practical lion dance programme were respectively aimed at knowledge acquisition, skills improvement and fulfilling personal interests, accounting for 61%, 58% and 57% of the surveyed students' respective responses. The results have reflected that outdoor training is highly popular and the students love it.

With regards to exploratory factor analysis, three common factors were obtained: practical curriculum value, practical curriculum development and practical curriculum teaching methods. Practical curriculum, its design and teaching methods were the three variables set as contributing factors. In our regression analysis, the Lingnan lion practical curriculum design set as the dependent variable as against the curriculum value and teaching approach as the independent variables, we have obtained a significant value for the regression coefficient of practical curriculum with strong influence on practical curriculum design (design factor = 2.263 + 0.635*practical curriculum value +0.163*practical curriculum teaching style).

(b) The practical curriculum is an effective way to disseminate the Lion Dance Intangible Cultural Heritage

It is found that the traditional master-apprentice model has lagged behind and has given way to a great extent to the newly practical-designed curriculum activities which are more proficient in training and mentoring students with professional skills. Through the observation and interviews, the program is found to be not only an important educational means to effectively mobilize students' interest in learning, but also help cultivate their team spirit, values of hard work and mutual tolerance. This can help further enhance China's self-confidence in the pride of its traditional heritage.

Through the authors' participatory observations, interviews with travel agency managers and a lion club president, teachers and students in Guangdong and Hong Kong unanimously recognized the value of the curriculum as the way to disseminate valuable cultural heritage. In this study, the interviewees believed that the program could help them understand the intangible culture of the Lingnan's Awakening Lion Dance and could help them profoundly experience the benefits of this lion dance culture of practising martial arts and its revering virtue and strength, *kung fu*, the beauty of the art of drumming, costumes and props, and the joy of customs and celebrations.

This paper demonstrates a Chinese case study which shows the educational advantage provided by organizing more outdoor learning activities that have given educators food for thought about the mutual reinforcement effect via knowledge acquisition and improvement in physical fitness backed up by practical lion dance exercise. Through cross-organizational research and analysis, this paper uses a combination of qualitative and quantitative approach for analysis and it argues in favour of the positive impact and values of the practical curriculum teaching methods, and the results have been well received by school teachers, students, travel agency managers, and a president of the lion dance community. This pedagogical experience well deserves further in-depth research and explorations.

Acknowledgement for Funding This project is sponsored under the "Special Research on Chaozhou Culture" (GD22CZZ01) with fund provided by the "Key Project of Philosophy and Social Science Planning of Guangdong Province", China.

References

Arcodia, C., Abreu Novais, M., Cavlek, N., & Humpe, A. (2021). Educational tourism and experiential learning: Students' perceptions of field trips. *Tourism Review, 76*(1), 241–254.

Avaunt, C. (2021). Impermeable bodies: Women who lion dance in Boston's Chinatown. *Theatre, Dance and Performance Training, 12*(2), 149–163.

Bachner, D., & Zeutschel, U. (2009). Long-term effects of international educational youth exchange. *Intercultural Education, 20*(sup1), S45–S58.

Cao, X. J. (2020). *The study on the construction of collective memory of Cantonese about "Guangdong lion"—Take the present situation of Guangzhou and Foshan Guangzhou*. Jinan University.

Cuffy, V., Tribe, J., & Airey, D. (2012). Lifelong learning for tourism. *Annals of Tourism Research, 39*(3), 1402–1424.

Davis, R., Bolden-Tiller, O., & Gurung, N. (2016). Experiential learning for Tuskegee University students: Study tour of agricultural industries. *Journal of Animal Science, 94*(suppl_1), 64.

Doğantan, D. E. (2020). An interactive instruction model design with role play technique in distance education: A case study in open education system. *Journal of Hospitality, Leisure, Sport & Tourism Education, 27*(1), 100268. https://doi.org/10.1016/j.jhlste.2020.100268

Dredge, D., Benckendorff, P., Day, M., Gross, M. J., Walo, M., Weeks, P., & Whitelaw, P. (2012). The philosophic practitioner and the curriculum space. *Annals of Tourism Research, 39*(4), 2154–2176.

Girvan, C., Conneely, C., & Tangney, B. (2016). Extending experiential learning in teacher professional development. *Teaching and Teacher Education, 58*, 129–139.

Guachalla, A., & Gledhill, M. (2019). Co-creating learning experiences to support student employability in travel and tourism. *Journal of Hospitality, Leisure, Sport & Tourism Education, 25*, 100210. https://doi.org/10.1016/j.jhlste.2019.100210

Howley, A., Howley, M., Camper, C., & Perko, H. (2011). Place-based education at Island community school. *The Journal of Environmental Education, 42*(4), 216–236.

Jick, T. D. (1979). Mixing qualitative and quantitative methods: Triangulation in action. *Administrative Science Quarterly, 24*(4), 602–611.

Kim, H. J., & Jeong, M. (2018). Research on hospitality and tourism education: Now and future. *Tourism Management Perspectives, 25*, 119–122.

Kolb, A. Y., & Kolb, D. A. (2009). The learning way: Meta-cognitive aspects of experiential learning. *Simulation & Gaming, 40*(3), 297–327.

Lee, J. (2004). *Managing educational tourism*: Brent W. Ritchie with N. Carr and C. Cooper (2003) (pp. xvii+283). Channel View Publications. ISBN 1-853150-51-2 (hbk); 1-853150-50-4 (pbk), *25*(5), 640–641.

Li, P., & Liang, H. (2020). Factors influencing learning effectiveness of educational travel: A case study in China. *Journal of Hospitality and Tourism Management, 42*, 141–152.

Ling, L. X. (2019). *The value of dragon and lion dance sports and its significance in colleges and universities*. Sports World: Academic Edition.

Liu, J., & Chen, C. (2020). The construction of a sustainable ecosystem for the study tour industry in the post-epidemic era. *Journal of Tourism, 35*(9), 4.

McGladdery, C. A., & Lubbe, B. A. (2017). International educational tourism: Does it foster global learning? A survey of South African high school learners. *Tourism Management, 62*, 292–301.

McGuire, C. (2019). War drums in Chinatown: Chinese Canadian lion dance percussion as martial art. In A. Hoefnagles, J. Klassen, & S. Johnson (Eds.), *Contemporary musical expressions in Canada* (pp. 242–274). McGill-Queen's University Press Montreal.

Olga, J. (2004). Book reviews. *Journal of Travel Research, 43*(1), 92–92.

Özoğul, T., Sezerel, H., Aktaş, S. G., & Adıgüzel, Ö. (2020). Learning through creative drama in vocational hospitality education: Efficiency, retention, and attitudes. *Tourism Management Perspectives, 36*, 100728.

Petroman, C., Mirea, A., Lozici, A., Constantin, E. C., Marin, D., & Merce, I. (2016). The rural educational tourism at the farm. *Procedia Economics and Finance, 39*, 88–93.

Riegel, U., & Kindermann, K. (2016). Why leave the classroom? How field trips to the church affect cognitive learning outcomes. *Learning and Instruction, 41*, 106–114.

Stone, M. J., & Petrick, J. F. (2013). The educational benefits of travel experiences: A literature review. *Journal of Travel Research, 52*(6), 731–744.

Thompson, M. A., Lee, S. S., Seegmiller, J., & McGowan, C. P. (2015). Kinematic and kinetic comparison of barefoot and shod running in mid/forefoot and rearfoot strike runners. *Gait & Posture, 41*(4), 957–959.

Tian, L. C., & Kong, Y. (2022). Curriculum design of integrative practice activities from the perspective of big idea. *Contemporary Educational Sciences, 01*, 23–30.

Tribe, J. (2002). Education for ethical tourism action. *Journal of Sustainable Tourism, 10*(4), 309–324.

Van Klooster, E., van Wijk, J., Go, F., & van Rekom, J. (2008). Educational travel: The overseas internship. *Annals of Tourism Research, 35*(3), 690–711.

Velempini, K., & Martin, B. (2019). Place-based education as a framework for tourism education in secondary schools: A case study from the Okavango Delta in Southern Africa. *Journal of Hospitality, Leisure, Sport & Tourism Education, 25*, 100197.

Xu, S. G., Zeng, Z. Q., & Chen, Y. H. (2017). Research on the function of dragon and lion dance under the background of the national fitness. *DEStech Transactions on Social Science, Education and Human Science*. https://doi.org/10.12783/dtssehs/eeres2016/7621

Yang, F. X., & Lau, V. M. C. (2019). Experiential learning for children at world heritage sites: The joint moderating effect of brand awareness and generation of Chinese family travelers. *Tourism Management, 72*, 1–11.

Ying, L. F., & Chiat, L. F. (2016). Dramatizing Malaysia in contemporary Chinese lion dance. *Asian Theatre Journal, 33*(1), 130–150.

Zeng, W. D. (2022). Research on the integration of study tour and comprehensive practical activity curriculum. *Gansu Education Research, 18*(10), 128–130.

Zhang, K., Wang, H., Wang, J., Li, C., & Zhang, J. (2021). Nature-based experiential learning as a framework for preparing responsible tourism practitioners. *Journal of Hospitality, Leisure, Sport & Tourism Education, 29*, 100297. https://www.researchgate.net/publication/347465589_Nature-based_experiential_learning_as_a_framework_for_preparing_responsible_tourism_practitioners

Zhou, D. (2018). The feasibility research of organizing dragon-and-lion dance in the ordinary universities of Wuhan. In *Proceedings of the 2018 3rd international conference on electrical, automation and mechanical.*

Xiaoyu Wang is a PhD candidate in tourism management at Jinan University. He has BA and MA degrees respectively from The Open University of China and South China Normal University. His main research interests are in hotel management, tourism economics, technology adoption in tourism, and the development and conservation of intangible cultural heritage. In addition to participating in a major project of the National Social Science Foundation, he has participated in the development of teaching materials. He has published nearly a dozen papers in national and international journals.

Zhen Guo (Eric) received his Bachelor's degree in tourism management from Qingdao University in China. He is currently pursuing a Master's degree in tourism management in Shenzhen Tourism College of Jinan University. His main research interests focus on intangible cultural heritage tourism development and management, leisure tourism. So far, he has published an article, titled *Online Reservation Intention of Tourist Attractions in the COVID-19 Context: An Extended Technology Acceptance Model*. In late 2022, he co-authored another paper, published in *A Bibliometric Analysis of Research on Intangible Cultural Heritage Tourism Using CiteSpace: The Perspective of China*.

Ziyi Qiu is a PhD candidate in Tourism Management at Jinan University, China. She has her BA and MA degrees from the Sichuan Normal University and Beijing International Studies University respectively. Her main research interests are in tourism destination marketing and destination management, and more recently heritage tourism. In addition to participating in a major project of the National Social Science Foundation, she has participated in the compilation of more than 10 books and published academic articles in journals, including *The Origin and Evolution of National Culture Park Management Policy* and *Reform of Cultural Heritage System Under the Background of National Cultural Park Construction*.

Mu Zhang received his BSc and MS degrees from Guangxi Normal College and Nanjing Normal University respectively, and his PhD from Fujian Normal University. Since December 2004, he has served as Associate Dean of Shenzhen Tourism College of Jinan University, Professor, Director of Tourism Experimental Teaching Center of Jinan University, and Chinese Director of the Joint Laboratory on IT & Tourism of Jinan University-University of Florida (JNU-UF) and International Joint Laboratory on IT & Tourism). He is a member of the Guangdong Province Tourism Management Teaching Steering Committee. He was a member of the Teaching Steering Committee of Tourism Management Specialties of Higher Education Institutions under the Ministry of Education, and his main research interest is the digital preservation of intangible cultural heritage. In addition to the textbooks and books he has edited, he has written more than 30 academic papers, many of which published in international journals.

Chapter 12
Transformative Tourism: Introducing the Experiential Cultural Tourism Model as an Innovative Means to Cultivate Resilient Ends for COFA Nations

James E. Faumuina

1 Introduction

In Thomas Pynchon's 'Gravity's Rainbow,' the explosive spectacle of missiles creates a vivid tapestry, painting the skies with a glorious, ominous glow (Pynchon, 1973). This literary masterpiece immerses readers in the chaotic aftermath of missile detonations, capturing the raw essence of historical moments charged with both destruction and an eerie aesthetic.

Drawing parallels to this evocative imagery, we find ourselves contemplating a different kind of glow—the lingering radiance from nuclear testing in the Bikini Atoll in the Marshall Islands. Just as Pynchon's narrative unveils the surreal brilliance of missile explosions, the Bikini Atoll witnessed its own display of destructive brilliance, albeit one tied to the somber legacy of nuclear experimentation (Ballendorf, 2005).

This radiant glow, born from the fusion of science and destruction, echoes through time, leaving behind a landscape rich with historical significance. In the aftermath of such events, the question arises: Can these sites become more than remnants of the past? Could they, like the explosive scenes in "Gravity's Rainbow," capture the observer's fascination with the detonation and cause them to embrace it as an ethereal existential guidepost? And when we think of the essence of tourism, could witnessing the generational damage done by this atomic legacy serve a higher purpose, can it be something beyond leisure and employed in a way that creates an opportunity for the tourist to become something more than a consumer?

These inquiries challenge the concept of tourism and test its utilitarian application as a genuine means to an end. This chapter deconstructs tourism in a way that conveys the most benefits to those who generally retain the least reward in the

J. E. Faumuina (✉)
Troy University, Troy, AL, USA
e-mail: jfaumuina@troy.edu

T.-C. Wong et al. (eds.), *Cultural Tourism in the Asia Pacific*,
https://doi.org/10.1007/978-3-031-63459-8_12

traditional tourism model. Transforming the old methodology of tourism in favor of a community asset-based approach could be the rebirth of an antiquated industry by focusing on the experiential cultural tourism model (ECTM) as a novel perspective on tourism for countries in the Pacific, namely those in Micronesia.

1.1 What Is the Experiential Cultural Tourism Model?

"Cultural tourism offers significant economic benefits to museums and heritage sites, but requires focusing on customer service, partnerships, and entrepreneurial approaches to increase attendance and self-generated revenues while maintaining heritage preservation and education mandate" (Silberberg, 1995: 361). There are many benefits of applying this economic tool to economically empower the Compact of Free Association (COFA) nations, made of three sovereign countries, the Federated States of Micronesia, the Republic of the Marshall Islands, and the Republic of Palau, including the merits of positioning them to become beneficiaries of experiential cultural tourism. Offering opportunities for tourists to learn and actively engage with the community about their nuclear heritage could transform tourism into an experiential cultural tourism model. This, in turn, could create a community asset-based commodity that summons tourists to engage in tourism (McGovern, 2023).

This chapter introduces two concepts developed for this model: The Resilient Intersectional Community-Based Empowerment (RICE) Framework, influenced by Conconi and Viollaz (2017) and the Community-Based Economic Development Wealth Equity Building Blocks (CBED WEBB), built on ideas from Theodos et al. (2021). These frameworks are synthesized into a cohesive model that creates the Experiential Cultural Tourism Model (ECTM). Functioning as an interdisciplinary approach, the ECTM integrates the principles from these concepts, channeling them into experiential engagements for tourists. Within this model, the heritage and lived experiences of the communities are not passive elements in the landscape, they transform into economic assets.

The ECTM strongly emphasizes engaging tourists not only as observers of history but also as active participants in the cultural tourism experience. It encourages meaningful interactions by facilitating encounters with survivors, activists, and community leaders during visits. Additionally, the model extends beyond traditional tourism by offering tourists opportunities to partake in environmental restoration and social empowerment initiatives. This approach is geared towards creating immersive and engaging experiences that go beyond the conventional tourist role.

The chapter continues with an example of practical applications of the ECTM to the COFA nations. This begins by exploring systemic poverty systems and understanding the profound consequences of disasters on disparate communities. This application involves interventions from intermediary resources, exploring community deficit (University of Memphis, 2023)and asset assessment, and transforming community assets in support of the ECTM.

To further contextualize the importance of ECTM, we turn to international frameworks. The United Nations World Tourism Organization (UNWTO) Framework Convention on Tourism Ethics underscores the imperative for tourism policies to respect and protect artistic, archaeological, and cultural heritage, ensuring accessibility in tourism. According to the UNWTO, cultural tourism is characterized by visitors' intrinsic motivation to learn, discover, and experience tangible and intangible cultural attractions/products, encompassing a society's material, intellectual, spiritual, and emotional features.

Exploring sustainable practices and cultural engagement, the UNWTO and the World Indigenous Tourism Alliance (WINTA) have compiled a Compendium of Good Practices in Indigenous Tourism. While not directly focused on COFA nations, this compendium offers insights into fostering applicable sustainable tourism practices and cultural preservation (WTO-WIT, 2023) .

1.2 *Purpose of the Experiential Cultural Tourism Model*

Integrating the RICE Framework and CBED WEBB as the ECTM and applying this to the COFA nations shifts the paradigm of an industry often seen merely as leisure-based to one where the opportunity for experiential engagement leaves a profound imprint on the traveler. By creating a commodity of community-based assets, they can leverage events such as the COFA nations'distinctive nuclear heritage and strategically address challenges posed by the global climate crisis.

Regarding COFA adapting ECTM, the emphasis remains fixed on community empowerment, equity, and cultural sensitivity. Additionally, it acknowledges the specific challenges these nations face in addressing systemic poverty systems and the threat of ever-present external destabilizing events such as natural disasters. This chapter introduces strategies and recommendations, all with a singular focus: improving the quality of life for COFA nations by creating a bridge for economic development, preserving cultural heritage, and fortifying community resilience by applying the ECTM to facilitate sustainable cultural tourism.

The ECTM initiates a paradigm shift with a focus on benefiting various shareholders (not stakeholders, due to this term's connotation of colonialism) involved in the development of the COFA nations. At the forefront are the local communities, where ECTM catalyzes economic empowerment, cultural heritage preservation, and enhanced community resilience. The next set of shareholders includes tourists, who are offered a unique and immersive cultural experience, transcending conventional tourism norms. Additional shareholders encompass the respective governments of COFA nations, managing the accumulated benefits of economic development, improved quality of life, and global recognition of the cultural significance of the COFA nations. Intermediary organizations play a vital role as shareholders in facilitating community capacity development. In essence, ECTM emerges as a holistic approach, intricately weaving together cultural preservation, economic development, and environmental sustainability, ensuring widespread benefits for all involved shareholders.

2 COFA Nation's Disparate Need for Sustainable Options

2.1 *Understanding Poverty Systems*

To grasp the intricacies involved in the COFA nations, an exploration of poverty systems, including the core livelihood systems of income maintenance, food security, health, education, and housing, needs to unfold. Studies highlight critical gaps in services that directly contribute to poverty alleviation within poor countries like the COFA nations (Suich et al., 2015). A sufficient understanding of community-based poverty alleviation techniques continues to elude research. In reviewing these studies, it underscores the imperative need for more nuanced analyses, demanding consideration of both direct and indirect impacts on poverty systems.

As articulated in a report by the U.S.Commission on Civil Rights (USCCR, 2019), COFA migrants face disproportionate challenges, including homelessness, unemployment, and inadequate healthcare. This is compounded by a lack of information in their language which contributes to systemic discrimination, impeding their access to decent housing, employment, and essential government services. Within these poverty systems lies a spectrum of challenges, ranging from restricted access to fundamental resources like education and healthcare to the inadequacy of critical infrastructure. This report reinforces this, emphasizing the struggles faced by COFA migrants in areas such as language barriers, employment, and housing.

Historical legacies, acculturation issues with the United States (Hagiwara et al., 2016), and debilitated infrastructure (WTO-WIT, 2023) contribute to the persistence of poverty within these systems. Unraveling these complexities lays a foundation for targeted interventions, challenging the visioning of sustainable development and the empowerment of COFA communities, and it is here where we start to explore how tourism can serve as a means to effect generational poverty's end.

In essence, the poverty systems, influenced by both environmental and socio-economic factors, paint a vivid picture of the challenges faced by COFA nations. Recognizing the intersectionality (Minear & Erevelles, 2013) of these challenges becomes imperative for crafting effective interventions (Shelton et al., 2021) such as the concept of ECTM as a community asset-based instrument for development and resilience.

2.2 *Background of COFA Nations*

The origins of the Compacts of Free Association trace back to the aftermath of World War II, when the United Nations entrusted the United States with administering authority over these strategically designated Pacific Islands as a safeguard for national security (Ballendorf, 2005). Secretary of State George Marshall, during the 1947 hearings to ratify the United Nations trusteeship, affirmed that the resolve of the American people was to prevent the use of this area as a springboard for aggression (USCCR, 2019).

However, the historical narrative takes a stark turn as between 1946 and 1958, the United States conducted at least 67 nuclear tests in the Marshall Islands, significantly impacting Micronesia and its people. The fallout from these tests forced residents to relocate, rendering traditional agriculture impossible on irradiated lands (USCCR, 2019). The economy became dependent on the United States, highlighting the human problems that arose amid the strategic security considerations.

The web of poverty systems casts a multifaceted layer on the COFA nations, echoing through various components of the community. Economic disparities, a byproduct of limited opportunities and uneven wealth distribution, give rise to a pervasive cycle of poverty within these communities. Education and healthcare, pillars essential for development, bear the weight of these internal poverty dynamics. Constraints imposed by these systems curtail access to quality education and healthcare services (Hagiwara et al., 2016), perpetuating a cycle of limited opportunities for future generations. Examining these impacts lays the groundwork for targeted interventions and the proposition that community-based assets exist and if cultivated they could be applied toward their resilience.

2.3 *Geo-Political Vulnerability of COFA Nations*

The vulnerability of COFA nations to external destabilizing events is closely tied to their existing dynamics. A disruption in one area, whether economic or environmental, can set off cascading effects (Wiwad et al., 2020). Recognizing the interconnected nature of these challenges is fundamental in considering how ECTM could foster economic development and resilience within COFA nations. External factors, including natural disasters, global economic shifts, and geopolitical influences, have the potential to heighten vulnerabilities within these communities (Mechanic & Tanner, 2007). These gains added significance against the backdrop of evolving dynamics in the Indo-Pacific region, a focal point for the United States (Biden, 2022).

In the Indo-Pacific region, the U.S. recognizes its economic significance, playing a pivotal role in supporting U.S. Regional Influence. The strategy underscores the importance of forming strong alliances and coalitions, highlighting partnerships with countries in the Pacific Islands that align with the geopolitical position of the U.S. and the COFA nations. The focus on resilience and security aligns with its exploration, seeking to enhance the ability of nations to weather external shocks, whether natural disasters or transnational actions.

Building upon this context provided on the COFA nations, it underscores the interconnected nature of challenges, examining vulnerabilities arising from both poverty systems and external destabilizing events. Coincidentally, this area of the world becomes particularly interesting for those seeking an experiential engagement through cultural tourism as a chance to be involved in historical events. Shaping the ECTM in the COFA nations offers experiential opportunities to witness historical changes, making it a compelling prospect for individuals interested in engaging with and understanding this unique geopolitical landscape.

3 Applying ECTM Means Toward COFA Nations

3.1 *RICE Framework and CBED WEBB Concept*

In today's dynamic global habitat, the challenges of addressing economic disparities, fostering community resilience, and achieving sustainable development often appear disconnected. This chapter will introduce an innovative and integrated approach by synergizing the Resilient Intersectional Community-Based Empowerment (RICE) Framework with the CBED WEBB (Community-Based Economic Development Wealth Equity Building Blocks) concept both combined as generators for transformation in the application of ECTM.

(a) *Resilient Intersectional Community-Based Empowerment (RICE) Framework*
In considering transforming ECTM, there needs to be a strategic approach that emphasizes inclusivity, empowerment, and community engagement to drive sustainable economic development in communities impacted by poverty and external destabilizing events (Conconi & Viollaz, 2017). The RICE Framework focuses on creating more sustainable, just, and inclusive societies, providing a roadmap for policymakers, community leaders, and advocates. Key components of this framework that have been adapted from the tenets of Community-Based.

Economic Development include resilient community building, developing intersectional perspectives, community-based approaches, empowerment and inclusivity, empowering marginalized communities, and lifting disenfranchised populations (Adaptation Fund Board, 2022).

(b) *CBED WEBB (Community-Based Economic Development Wealth Equity Building Blocks)*

This is a concept focusing on building the capacity to develop wealth in underdeveloped communities as a community-driven, equity-focused, and culturally sensitive strategy for economic development. What differentiates CBED WEBB from the RICE Framework is its focus on the development of wealth specifically as a means to attain equity. The advantages of both RICE and CBED WEBB are given in Table 12.1 below. In order to create wealth equity, there is a need to leverage local strengths and resources while respecting cultural and environmental considerations (CBED, 2021). Examples of these strategies include the development of microenterprises and small businesses, local cooperatives, skills training and education. Other measures also include social impact investing, cultural and heritage preservation, and digital inclusion and connectivity (Theodos et al., 2021).

3.2 *Transforming Tourism into an Experiential Engagement*

Given the challenges posed by poverty systems and external factors, COFA nations are compelled to explore alternative avenues for sustainable development. In this context, reshaping tourism into ECTM becomes not only an opportunity but a necessity.

Table 12.1 RICE and CBED WEBB advantages

Item	Description
Proposal	Advantages of synergizing resilient intersectional community-based empowerment (RICE) and CBED WEBB for transformative development
Abstract	This table explores the synergy between the resilient intersectional community-based empowerment (RICE) Framework and the CBED WEBB (community-economic development wealth equity building blocks) concept as a proposed strategy for driving sustainable economic development. By combining RICE's emphasis on inclusivity, empowerment, and community engagement with CBED WEBB's community-driven, equity-focused, and culturally sensitive approach, this model presents a comprehensive framework that addresses unique challenges and opportunities.
Introduction	Distinct economic challenges, including isolated economies, marginalized communities, and limited resources, necessitate a holistic approach. This integration of the RICE Framework and the CBED WEBB concept offers transformative solutions that can resonate in various contexts.
Advantages of integrating RICE and CBED WEBB	Inclusive community empowerment: The integration fosters a sense of ownership and participation among marginalized communities, promoting their active involvement in economic activities. Cultural sensitivity: This combined approach ensures that economic development initiatives are culturally relevant and inclusive, respecting diverse traditions and values. Resilient community building: The synergy helps communities withstand challenges and tap into their inherent strengths for sustainable growth. Customized solutions: This hybrid approach designs customized solutions, addressing unique challenges in innovative ways. Long-term sustainability: The integration places long-term sustainability at the forefront, creating a foundation for lasting economic growth and stability. Maximizing limited resources: By utilizing local resources and community-based approaches, this strategy presents a practical development model that doesn't heavily rely on foreign investment and commercial capital.

Sources: Author generated (Theodos et al., 2021; Conconi & Viollaz, 2017; CBED, 2021)

Transitioning to the discussion on external destabilizing events and the United States Indo-Pacific Strategy, the concept of transforming tourism into ECTM emerges as a viable option for COFA nations, driven by their unique circumstances and geographic location in the Indo-Pacific.

COFA nations possess a wealth of internal resources, and the transformation of tourism into ECTM aligns with the imperative to create assets from these resources. One distinctive asset lies in their nuclear legacy (USCCR, 2019), an aspect that can be reframed as a commodity for experiential tourism. The historical narrative of nuclear testing in the region, coupled with the resilience and adaptation of local communities, can be packaged as a compelling experiential journey for tourists. This approach not only leverages historical events but also fosters a sense of community empowerment by presenting their story as something that can be shared with a global audience (Fetterman, 2002).

Moreover, climate change preparation becomes another valuable aspect that can attract tourists. COFA nations, situated in vulnerable geographic locations, are at the forefront of climate-related challenges (Rasmussen et al., 2022). By showcasing their initiatives, adaptation strategies, and sustainable practices, these nations can position themselves as leaders in climate action. Tourists seeking immersive experiences aligned with environmental consciousness may find value in understanding how these communities navigate the impacts of climate change.

The transformation of tourism into the Experiential Cultural Tourism Model aligns with the global shift toward sustainable and responsible travel. This evolution offers the COFA nations a distinctive value proposition, portraying them not merely as tourist destinations but as educators, storytellers, and custodians of their unique narratives. Embracing their ancestral history as master seafarers, made popular by master navigator Mau Piailug, also known as "Papa Mau",[1] the COFA nations can use the ECTM to showcase their historic lineage as voyagers and navigators. However, navigating this dynamic shift into ECTM necessitates specialized assistance, which will be further explored through a third-party economic intermediary agent for the COFA nations.

3.3 *Value of Intervention from Intermediary Resources*

In the wake of the 2023 Maui Wildfire, the island of Maui was devastated by the loss of its economic generator—the town of Lahaina. Out of this event, the Hawaii Pacific Advisory Group emerges as a necessary intermediary, orchestrating a delicate dance between development and the preservation of the island's unique cultural and natural heritage that creates its highly valued sense of place. HPAG, a volunteer-driven organization, uniquely positioned to navigate the complexities inherent in sustaining the local economy and specifically ensuring that a thriving tourism industry would endure in the recovery phase, all the while ensuring it would remain responsive to the feedback it received from the local community (HPAG, n.d.).

HPAG's role in Maui could mirror the type of economic intervention required from an intermediary in the COFA nations. This extends beyond mere facilitation; it embodies the principles of coordination, responsiveness, and collaboration (Pfadenhauer et al., 2017) Drawing from its volunteer members who have years of experience in addressing the immediate needs of communities affected by disasters, HPAG operates with a feedback-oriented and collaborative approach (Gilmartin et al., 2018) as an intermediary, an organization like HPAG would navigate the challenges faced by COFA nations, striving to strike a balance between economic development and the imperative to conserve their distinct heritage.

[1] The Revival of Wayfinding in the Pacific—Mau's influence inspired a revival of traditional wayfinding across Polynesia, Micronesia, and Melanesia.

(a) *Contributions to Economic Development*

The intervention of an intermediary organization like HPAG with the COFA nations meaningfully contributes to economic development by unlocking the untapped potential of their cultural and natural assets. Through strategic partnerships and investment facilitation, an intermediary organization would bring in external resources and expertise, propelling the development of sustainable initiatives like the ECTM.

In context, considering cultivating an area like the Bikini Atoll for ECTM, the involvement of an intermediary organization could prove pivotal in attracting shareholders, supporters, and resource investors aligned with the ambitious objective of transforming cultural tourism for the COFA Nations. In leveraging the islands' distinctive nuclear legacy and climate change challenges, such an intermediary organization would strategically position and brand the COFA nations as areas to visit in order to engage with the pioneers in disaster resilience and climate change.

Additionally, incorporating an equity-focused perspective as a planning function of the intermediary that is supporting the ECTM model ensures that economic benefits are distributed fairly among its community members, including the disenfranchised and disabled. This aligns with the need for community empowerment that is addressed in the RICE and CBED WEBB concepts, and it also contests the economic disparities of countries in need of vital economic development.

Economic diversification is another facet where an intermediary can provide useful consultation to the COFA Nations if they were to pursue the ECTM. By identifying and promoting alternative sources of revenue from ECTM, such as eco-tourism, cultural exchanges, and community-based enterprises, an intermediary like the HPAG could facilitate capacity building.

(b) *Facilitating Collaboration and Resource Mobilization*

Intermediary organizations serve as catalysts for collaboration and resource mobilization, essential in securing the resources needed to transform tourism into the ECTM. Through structured facilitation, these types of organizations would be well-equipped to facilitate partnerships between COFA nations and various shareholders, including international organizations, NGOs, and private investors (Pfadenhauer et al., 2017).

Intermediaries foster collaboration between local communities and external shareholders. By establishing platforms for dialogue and partnership building, these entities would enable COFA nations to access funding, technology, and knowledge that might be challenging to acquire independently. This collaborative approach would ensure that the ECTM initiatives are inclusive in design and respect the cultural nuances of the communities involved (Gilmartin et al., 2018).

(c) *Community Empowerment and Equity*

The overarching link to community empowerment is perhaps the most significant value that an intermediary organization could bring to COFA nations if they were to adopt the ECTM. By actively involving local communities in decision-making processes and development initiatives, intermediaries empower these communities to shape their destinies and create their social capital (Putnam, 2000).

Organizations like HPAG could potentially play a pivotal role in empowering communities through skill-building programs, educational initiatives, and promoting sustainable practices that could all feed into the ECTM as the workforce becomes upskilled in tandem with a new type of tourism industry being created.

An intermediary organization acts as a safeguard for preserving cultural heritage in the ECTM. Through their guidance, COFA nations can strategically integrate cultural preservation into economic activities, creating a synergy where the community's economic development aligns with safeguarding its unique traditions that become the forefront of tourist engagement in the ECTM.

To ensure the ECTM is equity-based, an intermediary organization would emphasize an Equity-Focused Quality Improvement (EFQI) approach (Weiner et al., 2022), involving equity assessment (Beck et al., 2020), community engagement (Glandon et al., 2017), and data-driven decision-making (Thompson, 2021). This would ensure that the ECTM initiatives are rooted in principles of justice and inclusivity (Pfadenhauer et al., 2017).

The potential intervention of intermediary organizations like HPAG in COFA nations is a strategic imperative for realizing the transformative potential of the RICE Framework and CBED WEBB concept. Their multifaceted role is assisting host nations in navigating challenges, facilitating collaboration, and empowering community organizations. They ensure economic initiatives like the ECTM become drivers of sustainable and inclusive development for these unique island nations considering implementing it as an economic resilience strategy.

3.4 ECTM Transformed as a Community Asset

(a) *Community Asset Assessment Process*

A "community asset"refers to any resource, attribute, or feature within a community that holds value, contributes to its well-being, and can be leveraged towards resilience, wealth, and sustainable development. In the context of ECTM, community assets encompass both tangible and intangible elements that play a role in developing the experiential experience for tourists. Their assessment plays a pivotal role in considering applying the ECTM. For the COFA nations, the assessment process involves an examination of which assets are present and which are lacking. This assessment is needed to ascertain how these community assets could be utilized to develop, support, and sustain this strategy. Grounded in the historical, socio-economic, and environmental context implicit in issues impacting poverty in the COFA nations, this chapter will attempt to steer through the assets and deficits in the communities, as it aims to navigate these complexities by identifying and understanding deficits and drawing insights from them (Fisher et al., 2014).

Embracing RICE and CBED concepts that underscore the significance of identifying and leveraging cultural and natural assets, this chapter discusses why it is necessary to have an evaluation of these assets and deficits (University of Memphis, 2023) as the components of the assessment. It envisions experiential tourism as a strategic avenue for fostering sustainable community development. It attempts to connect the proposed interventions with the theoretical frameworks and align them with the distinct needs and assets of COFA communities, paving the way for a sustainable and innovative development trajectory (Mastroberardino et al., 2022).

(b) *Identifying Deficits in Communities*

The process of identifying service deficits within COFA communities stands as a pivotal step, essential for formulating targeted and effective interventions that address the nuanced challenges faced by these unique communities. This stage is grounded in the exploration of the historical, socio-economic, and environmental context presented in the previous sections, serving as a discerning lens through which we examine the multifaceted dimensions of deficits and disparities (Fisher et al., 2014).

This approach ensures that identifying deficits is not a one-size-fits-all endeavor but rather a tailored assessment that considers the specific context of the COFA communities. The deficits include various aspects, ranging from socio-economic disparities and healthcare limitations to educational constraints and infrastructure challenges (Ballendorf, 2005). The identification process is not solely about recognizing what is lacking but also understanding the interconnectedness of these deficits and their impact on the overall well-being of the communities (University of Memphis, 2023).

Drawing parallels from Beck et al's. (2020) examination of racial disparities in preterm birth outcomes, we recognize the value of developing assessments to garner community needs and deficits. Similarly, in the context of COFA communities, the acknowledgment of disproportionate challenges faced by COFA migrants adds another layer of complexity to the design of interventions (Hagiwara et al., 2016). This recognition ensures that our interventions are inclusive and considerate of the diversity within COFA communities.

An article by Civil Beat on President Biden's plan to restore benefits for COFA migrants (Grube, 2023) provides insights into the challenges and costs associated with providing social services to COFA migrants in the U.S. and its territories. The GAO report on compact migration estimates the high financial impact of COFA migrants on U.S. areas and discusses federal grants to defray the associated costs (U.S. Government Accountability Office, 2020). The assessment of these costs is a necessary component of ascertaining the community's capacity and ability to implement the ECTM.

Acknowledging the disproportionate challenges faced by specific demographic groups, there needs to be a means to have inclusive interventions within COFA communities to ensure the efficacy of the proposed ECTM endeavor. Reflecting on sources such as the USAID testimony on renewed U.S. engagement in the Pacific (USAID, 2017) and the ICMA article on facilitating inclusive community outreach and engagement (Trent, 2021) provides practical tips and examples for ensuring that inclusion is incorporated in the design of ECTM. The UNHCR provides an example of how this could look on its website where it discusses issues and initiatives supporting the development of welcoming the communities (UNHCR, n.d).

Navigating through this process, insights from constructs like the RICE Framework and CBED WEBB concept are presented to guide our understanding of deficits not merely as shortcomings but as opportunities for targeted resources toward areas that can be empowered and developed (Fisher et al., 2014). The goal is to transform deficits into focal points for strategic interventions that leverage community assets, cultural heritage, and environmental resources. Just as Beck et al. (2020) advocated for community engagement to address disparities and improve care, this chapter suggests that interventions for COFA nations are not only informed by theory but also rooted in feedback provided by those who live in these unique island communities.

(c) *ECTM incorporating the RICE and CBED WEBB*

Building upon the theoretical foundations presented with RICE and CBED WEBB, we can embark on the task of applying them to assess deficits and assets within COFA communities.

The intent in focusing on RICE and CBED WEBB is to explore how best to develop ECTM with inclusivity, cultural sensitivity, and community-based long-term sustainability. It reflects the principles of RICE and CBED WEBB, emphasizing community empowerment and customization of solutions tailored to the unique needs of each COFA community. This strategic application aligns with the overarching goal of maximizing limited resources and fostering a collaborative approach involving community members, local organizations, and external shareholders.

This process serves as a catalyst for forging a sustainable and community-driven path in transformative tourism development. By identifying and strategically leveraging these experiential assets, it lays the foundation for an approach that not only enhances the allure of COFA nations as tourist destinations but also has the potential to enrich the socio-economic fabric of the communities themselves by making ECTM community-based. In doing so, the ECTM has the potential to instill what Putnam (2000) prescribed as social capital, something that has proved to be a key social ingredient in thriving communities .

Table 12.2 illustrates the application of the RICE and CBED WEBB methods for two potential assets of the COFA nations, its nuclear heritage and its climate change adaptation. It highlights how a transition from conventional tourism to a more sustainable, inclusive, and culturally sensitive approach could occur.

4 Designing and Evaluating the ECTM

4.1 Crafting Sustainable Experiential Cultural Tourism Initiatives

Having explored the intricacies of poverty systems, shed light on the distinctive challenges faced by COFA nations, and addressed responses to their needs and service deficits, all the components are in place to assess the feasibility of the ECTM

Table 12.2 RICE and CBED WEBB criteria applied to ECTM

Item	Description
Introduction to the concept	Within sustainable community development, this section focuses on leveraging cultural and natural assets for experiential cultural tourism in COFA nations, aligning with global sustainable travel trends.
Strategic utilization of cultural assets	These methods guide the identification and strategic use of cultural assets like traditional maritime navigation practices and historical sites to create authentic cultural experiences.
Assessing and enhancing cultural engagement	The assessment process, rooted in the RICE framework, evaluates the potential for cultural engagement within experiential tourism, ensuring the recognition of cultural nuances.
Customized experiences for sustainability	Customization tailors experiential tourism offerings to meet the specific needs and context of each COFA community, fostering sustainability.
Long-term sustainability and collaboration	This section considers economic, social, and environmental impacts, involving collaboration with community members and external shareholders.
Maximizing limited resources for impact	The RICE and CBED WEBB methods consider the strategic use of limited resources, maximizing the value of cultural and natural assets to create impactful and meaningful tourism experiences.
Community empowerment through applying the concept	The ECTM initiative becomes a catalyst for inclusive community empowerment, creating economic opportunities and instilling pride in preserving cultural heritage.

Source: Author created

as an economic alternative initiative to traditional tourism and its potential as a community-based asset development strategy.

Here are the key criteria for the application of the RICE and CBED WEBB in the ECTM strategy:

1. Introduction to the Concept
2. Strategic Utilization of Cultural Assets
3. Assessing and Enhancing Cultural Engagement
4. Customized Experiences for Sustainability
5. Long-Term Sustainability and Collaboration
6. Maximizing Limited Resources for Impact
7. Community Empowerment through applying the Concept

In the pursuit of developing culturally appropriate tourism products for COFA nations, the active involvement of local communities in the product development process is necessary. Using the RICE and CBED WEBB to tailor its application to the unique contexts of nuclear heritage and climate change management, it could emphasize the significance of community engagement and empowerment in the following ways.

(a) *Introduction to the Concept*

The concept of experiential cultural tourism is introduced, emphasizing its distinction from other forms of tourism as the ECTM or "Experiential Cultural Tourism Model." Tourists are educated about the nuclear legacy of the COFA nations, creating awareness of its impact on culture, environment, health, and the community's resilience in the face of climate change.

(b) *Strategic Utilization of Cultural Assets*

Identification and strategic use of cultural assets, such as history, traditions, arts, and values, are guided by the RICE and CBED WEBB, ensuring a thoughtful and respectful approach. Stories, artifacts, and monuments related to nuclear testing and resistance movements are showcased to provide authentic cultural experiences that align with the community's narrative.

(c) *Assessing and Enhancing Cultural Engagement*

This method evaluates the level and quality of engagement between tourists and local communities, ensuring meaningful interactions. Tourists are encouraged to interact with survivors, activists, and leaders of nuclear justice campaigns to deepen their understanding of the community's history and struggles.

(d) *Customized Experiences for Sustainability*

Personalized experiences aligned with sustainability goals are designed, considering a focus on environmental and social responsibility. Tourists are provided with opportunities to participate in environmental restoration and social empowerment initiatives, fostering a sense of contribution.

(e) *Long-Term Sustainability and Collaboration*

The sustainability of the ECTM model is ensured by fostering collaboration among COFA nation development participants, including local communities, businesses, and government entities. Partnerships and alliances are established to support nuclear advocacy and awareness efforts in climate consciousness and adaptation, creating opportunities in the future for tourists to revisit COFA nations.

(f) *Maximizing Limited Resources for Impact*

RICE and CBED WEBB emphasize the strategic use of limited resources to maximize their responsible impact on the environment and community. Existing infrastructure and facilities are utilized efficiently to create engaging experiences for tourists, promoting resource optimization and providing them access to areas where they can firsthand observe the effect of the nuclear test that occurred and climate change currently taking place.

(g) *Community Empowerment through the Concept*

Local communities are empowered to participate in and benefit from the ECTM. The framework inspires communities to share their stories, perspectives, and aspirations, fostering a sense of identity and pride while actively engaging with tourists. This melding of cultures from the perspectives of both the tourist and local community enhances opportunities for social capital (Putnam, 2000) to be developed and heightens the probability of returned visits for those tourists who participate in ECTM.

4.2 Showcasing Cultural Heritage Responsibly

Transforming into ECTM, a course is set to redefine cultural tourism in COFA nations. Forefront is the responsible showcasing of their cultural heritage. Our vision will be implemented with the RICE Framework and CBED WEBB concept, as it aims not only to attract visitors but also, more significantly, to empower local communities. This approach becomes our guiding mantra, carefully balancing the exhibition of cultural richness with the preservation of local traditions.

The success of our experiential cultural tourism initiative would go beyond economic indicators. This commitment to community empowerment, preservation of cultural heritage, and equitable distribution of benefits is complemented by collaboration with local research institutions, Non-Governmental Organizations (NGOs) and community leaders to weave together a comprehensive transdisciplinary approach. For instance, applying this method to the unique contexts of nuclear heritage and climate change management as ECTM initiatives could bring forth several benefits, addressing key aspects.

Cultural Sensitivity: Utilizing community-based methodologies ensures a culturally sensitive development process, acknowledging the historical significance and cultural nuances of the COFA nations.

Inclusivity and Engagement: Actively involving local communities in the product development process aligns with principles of inclusivity, fostering genuine engagement between tourists and residents.

Sustainability: The focus on long-term sustainability, a core principle of community resilience strategies, ensures that tourism initiatives contribute positively to the environment, society, and economy.

Authentic Experiences: This distinctive feature sets ECTM apart from other forms of cultural tourism by prioritizing not only the tourist experience but also ensuring that these experiences contribute to the long-term sustainability of the community and align with cultural values and sensitivity. ECTM goes beyond generic encounters, reaching into the unique narratives of the community, and fostering a direct connection between tourists and the community.

Community Empowerment: The significance of empowering local communities extends across cultural preservation, economic vitality, and cultivating pride and resilience. This involvement contributes to the creation of economic opportunities as the benefits of tourism directly impact the well-being of residents, fostering a more equitable distribution of resources. It not only enhances the cultural experience but also fosters resilience, equipping communities with the capacity to navigate challenges while preserving their unique identity.

Educational Component: The introduction of tourists to the nuclear legacy or climate change challenges is relevant due to its emphasis on education. Ideally, the educational component not only enriches the tourist experience but also contributes to the broader objectives of sustainable cultural tourism by transforming tourists into COFA ambassadors who actively support the well-being and preservation efforts on behalf of the COFA nations to the world outside.

Collaborative Partnerships: By prioritizing collaborative partnerships, ECTM introduces a distinctive approach to cultural tourism. This emphasis on collaboration ensures that various shareholders are actively involved in shaping and sustaining tourism initiatives. This collaborative aspect fosters a more inclusive, comprehensive, and sustainable development process, acknowledging the uniqueness of the sense of place and the value of community-based assets. ECTM signifies a departure from conventional tourism models by promoting responsible and community-oriented practices that first prioritize the well-being of local communities, businesses, and cultural preservation efforts.

Integrating these elements, this proposed approach seeks to transform cultural tourism into a positive force for sustainable development, community empowerment, and responsible engagement with the cultural and natural heritage of the COFA nations. This structured framework provides a systematic way to address the complexities of tourism development while ensuring ethical, inclusive, and impactful outcomes.

4.3 Measuring Economic and Social Impact

Evaluating for effectiveness is not just a measure of success but a strategic and adaptive process that ensures the ECTM plan remains responsive, impactful, and aligned with the evolving needs of the COFA nations. It is an integral part of responsible and accountable project management, contributing to the long-term success and sustainability of the initiatives. By integrating these criteria into our evaluation strategy, we aim to not only measure the success of the ECTM plan but also understand and address the organizational readiness within COFA nations. This comprehensive approach reflects our commitment to ensuring a successful and contextually sensitive implementation of ECTM.

Table 12.3 below illustrates the role of effectiveness evaluation in the execution of the ECTM initiative for the COFA Nations. In aligning with Weiner's theoretical framework on organizational readiness for change (Weiner et al., 2022) and drawing insights from Pronovost and Marsteller's (2014) work on a fractal-based quality management infrastructure, our strategy for assessing the effectiveness of the ECTM plan is enabled with the following criteria as provided:

The evaluation of the ECTM serves as a multifaceted tool with several essential purposes. Firstly, it ensures accountability by holding shareholders responsible for the resources, efforts, and commitments invested in the ECTM plan. Secondly, the

Table 12.3 Evaluating the experiential cultural tourism model

Purpose	Description
Accountability	Effectiveness evaluation holds shareholders accountable for the resources, efforts, and commitments invested in the ECTM plan. It assesses whether intended outcomes align with initial goals and objectives.
Learning and improvement	Evaluation results offer insights into what works well and areas for improvement. It serves as a learning opportunity to refine strategies, adjust approaches, and enhance the overall plan that improve future outcomes.
Resource optimization	By assessing effectiveness, organizations identify areas where resources are optimally utilized and where adjustments are needed. This maximizes the impact of available resources and ensures efficient allocation in subsequent phases.
Shareholder communication	Effectiveness evaluation provides a basis for transparent communication with shareholders, including community members, investors, and various shareholders.
Adaptation to changing circumstances	Evaluation allows organizations to adapt to dynamic external changes, ensuring the continued relevance and effectiveness of the ECTM plan in response to evolving circumstances.
Demonstrating impact	Showcasing the positive impact of the ECTM plan on the community, economy, and environment through evaluation. This is important for gaining support, attracting investment, and promoting the sustainability of the initiatives.
Strategic decision-making	Evaluation results contribute to informed decision-making. Organizations can use findings to make strategic choices about scaling up, modifying, or discontinuing specific components based on observed impact.
Continuous improvement	An ongoing evaluation process supports a culture of continuous improvement, encouraging shareholders to seek innovative solutions, experiment with new ideas, and evolve strategies over time for enhanced effectiveness.

Source: Author created

evaluation process becomes a platform for learning and improvement, offering valuable insights to refine strategies and enhance the overall plan for better outcomes in the future. Resource optimization is another significant aspect, as organizations can identify areas where resources are utilized and areas that require adjustments.

Moreover, the evaluation process facilitates transparent communication with shareholders, demonstrates a commitment to openness and accountability, and fosters trust and support. It also enables adaptation to change circumstances, ensuring the continued relevance and effectiveness of the ECTM in response to evolving conditions. Additionally, evaluation serves to showcase the positive impact of the ECTM on the community, economy, and environment, crucial for gaining support and promoting sustainability. It contributes to strategic decision-making, supports continuous improvement, and ultimately ensures that the ECTM fulfills its intended objectives and positively contributes to the community.

5 Conclusion

In conclusion, this chapter attempts to navigate the complexities of cultural tourism in COFA nations, advocating for a paradigm shift from traditional models to the transformational ECTM employing constructs like the Resilient Intersectional Community-Based Empowerment (RICE) Framework and Community-Based Economic Development Wealth Equity Building Blocks (CBED WEBB) concept.

The exploration begins by contrasting historical narratives with contemporary challenges, weaving a concept that considers the application of historical sites for community empowerment and the ECTM. The advantages of are expounded upon, showcasing the transformative potential of inclusive community empowerment, community capacity building, and the development of resources. Intermediary organizations, exemplified by organizations like the Hawaii Pacific Advisory Group, are positioned as catalysts for sustainable tourism, balancing development with the preservation and promotion of the host nation's culture.

Giving action to theory, the chapter delineates a strategic process for deficit and asset assessment, deliberately transitioning to the development of a culturally sensitive and sustainable tourism initiative. Active involvement of local communities, responsible cultural heritage showcasing, and meticulous measurement of economic and social impact emerge as focal points.

The call to action is an urgent plea to transcend traditional paradigms and embrace the option to have transformative change by implementing global leadership through innovative initiatives like the ECTM. Just as Thomas Pynchon's 'Gravity's Rainbow' navigated the paradox of manmade destruction and the natural aesthetic, so too can COFA nations harness the dastardly acts placed upon them that caused an explosion over their historical horizon, converting it into something they now control as their own community assets and a catalyst for change. The integration of RICE and CBED WEBB, rooted in community empowerment and cultural sensitivity, provides a blueprint for crafting a sustainable tourism narrative. This call to action urges shareholders to collaborate, invest responsibly, and support intermediary organizations in realizing the vision of experiential engagement owned by the community. In embracing this call, COFA nations can embark on a journey where the echoes of past nuclear legacies resonate not only as reminders of destruction but as opportunities for cultural rebirth and resilience.

References

Adaptation Fund Board. (2022). *A study on intersectional approaches to gender mainstreaming in adaptation-relevant intervention*. https://www.adaptation-fund.org/wp-content/uploads/2022/02/AF-Final-Version_clean16Feb2022.pdf

Ballendorf, D. A. (2005). *Handbook of federal countries*. McGill-Queen's University Press.

Beck, A. F., Edwards, E. M., Horbar, J. D., Howell, E. A., McCormick, M. C., & Pursley, D. M. (2020). The color of health: How racism, segregation, and inequality affect the health and well-being of preterm infants and their families. *Pediatric Research, 87*(2), 227–234.

Biden, J. (2022). *Fact sheet: Indo-Pacific strategy of the United States*. https://www.whitehouse.gov/briefing-room/speeches-remarks/2022/02/11/fact-sheet-indo-pacific-strategy-of-the-united-states/

CBED (Community-Based Economic Development). (2021). *Hawaii community-based economic development technical and financial assistance program*. FY 2021 Annual & Financial Reports to the 31st Legislature. Legislative Report. Honolulu.

Conconi, A., & Viollaz, M. (2017). Development, poverty, and inequality: Changing the development paradigm. In *The age of perplexity: Rethinking the world we knew*. Oxford Poverty and Human Development Initiative. https://www.bbvaopenmind.com/en/articles/poverty-inequality-and-development-a-discussion-from-the-capability-approach-s-framework/

Fetterman, D. M. (2002). 2001 invited address: Empowerment evaluation: Building communities of practice and a culture of learning. *American Journal of Community Psychology, 30*, 89–102. https://doi.org/10.1023/A:1014324218388

Fisher, J. A., Patenaude, G., Giri, K., Lewis, K., Meir, P., Pinho, P., Rounsevell, M. D. A., & Williams, M. (2014). Understanding the relationships between ecosystem services and poverty alleviation: A conceptual framework. *Ecosystem Services, 7*, 34–45. https://doi.org/10.1016/j.ecoser.2013.08.002

Gilmartin, H., Lawrence, E., Leonard, C., McCreight, M., Kelley, L., Lippmann, B., & Burke, R. E. (2018). Brainwriting premortem: An efficient method for identifying implementation barriers. *Journal of Nursing Care Quality, 34*(2), 94–100.

Glandon, D., Paina, L., Alonge, O., Peters, D. H., & Bennett, S. (2017). 10 best resources for community engagement in implementation research. *Johns Hopkins Bloomberg School of Public Health, 32*(10), 1457–1465.

Grube, N. (2023). *Biden backs Hawaii's push to restore benefits to Micronesian migrants*. Honolulu Civil Beat. Dated April 11. https://www.civilbeat.org/2023/04/biden-backs-plan-to-restore-benefits-for-cofa-migrants

Hagiwara, M. K. I., Miyamura, J., Yamada, S., & Sentell, T. (2016). Younger and sicker: Comparing Micronesians to other ethnicities in Hawaii. *American Journal of Public Health, 106*(3), 485–491. https://doi.org/10.2105/AJPH.2015.302921

HPAG. (n.d.). *The Hawaii Pacific Advisory Group*. https://imagine-pacific.com/hpag.html

Mastroberardino, P., Calabrese, G., Cortese, F., & Petracca, M. (2022). New perspectives of experiential tourism: An exploratory analysis of live virtual tours during the COVID-19 outbreak. *The TQM Journal, 34*, 1732–1751. https://doi.org/10.1108/TQM-06-2021-0164

McGovern, C. (2023). Can tourism help recovery after a disaster? *National Geographic*. November issue. https://www.nationalgeographic.com/travel/article/how-can-tourism-help-recovery-after-disaster.

Mechanic, D., & Tanner, J. (2007). Vulnerable people, groups, and populations: Societal view. *Health Affairs, 26*(5), 1220–1230. https://doi.org/10.1377/hlthaff.26.5.1220

Minear, N., & Erevelles, A. (2013). Unspeakable offenses: Untangling race and disability. In J. Davis (Ed.), *The disability studies reader* (pp. 354–368). Routledge.

Pfadenhauer, L. M., Gerhardus, A., Mozygemba, K., Lysdahl, K. B., Booth, A., Hofmann, B., Wahlster, P., Polus, S., Burns, J., Brereton, L., & Rehfuess, E. (2017). Making sense of complexity in context and implementation: The context and implementation of complex interventions (CICI) framework. *Implementation Science, 12*(1), 21. https://doi.org/10.1186/s13012-017-0552-5

Pronovost, P. J., & Marsteller, J. A. (2014). Creating a fractal-based quality management infrastructure. *Journal of Health Organization and Management, 28*(4), 576–586. https://doi.org/10.1108/jhom-11-2013-0262

Putnam, R. D. (2000). *Bowling alone: The collapse and revival of American community*. Simon & Schuster.

Pynchon, T. (1973). *Gravity's rainbow*. Viking Press.

Rasmussen, E., Nakano, G., Kabua, S., & Hatoum, A. (2022). *Climate-focused field research within the Kwajalein Atoll Sustainability Laboratory* (Conference Paper: 465–468). https://doi.org/10.1109/GHTC55712.2022.9911040

Shelton, R. C., Adsul, P., Oh, A., Moise, N., & Griffith, D. M. (2021). Application of an antiracism lens in the field of implementation science (IS): Recommendations for reframing implementation research with a focus on justice and racial equity. *Implementation Research and Practice, 2*, 1–19. https://doi.org/10.1177/26334895211049482

Silberberg, T. (1995). Cultural tourism and business opportunities for museums and heritage sites. *Tourism Management, 16*(5), 361–365. https://doi.org/10.1016/0261-5177(95)00039-Q

Suich, H., Howe, C., & Mace, G. (2015). Ecosystem services and poverty alleviation: A review of the empirical links. *Ecosystem Services, 12*, 137–147.

Theodos, B., Marx, R., & Nunna, T. (2021). *Community wealth-building models: Opportunities to bolster local ownership*. Dated December, Urban Institute https://www.urban.org/sites/default/files/publication/105230/community-wealth-building-models_1.pdf

Thompson, V. (2021). *Understanding the policing of black, disabled bodies*. Center For American Progress. Dated February 10. https://www.americanprogress.org/article/understanding-policing-black-disabled-bodies/

Trent, S. (2021). How to facilitate inclusive community outreach and engagement. ICMA. *PM Magazine*. Dated May 1. https://icma.org/articles/pm-magazine/how-facilitate-inclusive-community-outreach-and-engagement

U.S. Government Accountability Office. (2020, June). *Compacts of free association: Populations in U.S. areas have grown, with varying reported effects*. Committee Report. Washington, DC.

UNHCR. (n.d.). *Promoting welcoming and inclusive societies*. The UN Refugee Agency https://www.unhcr.org/handbooks/ih/welcoming-inclusive-societies/promoting-welcoming-and-inclusive-societies

University of Memphis. (2023). *Comparison between asset and deficit based approaches*. Engaged Scholarship Curriculum https://www.memphis.edu/ess/module4/page3.php

USAID (U.S. Agency for International Development). (2017). Statement of Gloria Steele, Acting Assistant Administrator, Bureau for Asia, before the House Foreign Affairs, Subcommittee on Asia and the Pacific. Dated July 27. https://2017-2020.usaid.gov/news-information/congressional-testimony/jul-7-2017-gloria-steele-acting-aa-asia-hcfa-us-interests-asia-pacific-fy2018

USCCR (U.S. Commission on Civil Rights). (2019). *Micronesians in Hawaii: Migrant group faces barriers to equal opportunity*. Commission Report. Washington, DC.

Weiner, B. J., Lewis, C. C., & Sherr, K. (2022). *Practical implementation science: Moving evidence into action*. Springer.

Wiwad, D., Mercier, B., Piff, P., Shariff, A., & Aknin, L. (2020). Recognizing the impact of COVID-19 on the poor alters attitudes towards poverty and inequality. *Journal of Experimental Social Psychology, 93*, 104083. https://doi.org/10.1016/j.jesp.2020.104083

WTO-WIT (World Tourism Organization and World Indigenous Tourism). (2023). *Compendium of good practices in indigenous tourism—Regional focus on the Americas*. World Tourism Organization. https://www.e-unwto.org/doi/epdf/10.18111/9789284424184

James E. Faumuina is a Ph.D. student in the Troy University, Alabama, USA, Global Leadership Ph.D. program and a University Center of Excellence for Developmental Disabilities Trainee at the University of Hawaii Manoa's Graduate Program for Disability and Diversity Studies. He earned his BA in Psychology in 1993 and Master's degrees in Public Administration and Business Administration from Troy and Chaminade Universities respectively in 1996 and 2003. During his military service, he engaged internationally in delegations to Indonesia, Nepal, Philippines, Malaysia, and Guam and he has diverse professional experience, including leadership positions at The Leeward YMCA, Hawaii National Guard, and the Office of Hawaiian Affairs. He is serving as the Assistant Director of the Hawaii Pacific Advisory Group, owner of Imagine Pacific Enterprises and the author of the books *Triggers of Aggression in Crowds: The Milieu of High School Football* (2021) and *The Trial of Supercrip (And the Conviction of Narrative Prosthesis)* (2023).

Index

T.-C. Wong et al. (eds.), *Cultural Tourism in the Asia Pacific*, https://doi.org/10.1007/978-3-031-63459-8

R

S

T

Printed in the USA
CPSIA information can be obtained
at www.ICGtesting.com
CBHW071413250924
14770CB00009B/210

9 783031 634581